Anonymous

Letters and Papers on Agriculture and Planting, etc

Vol. I

Anonymous

Letters and Papers on Agriculture and Planting, etc
Vol. I

ISBN/EAN: 9783337108533

Printed in Europe, USA, Canada, Australia, Japan

Cover: Foto ©berggeist007 / pixelio.de

More available books at **www.hansebooks.com**

LETTERS AND PAPERS
ON
Agriculture, Planting, &c.

SELECTED FROM

THE CORRESPONDENCE

OF THE

Bath and West of England Society

FOR THE ENCOURAGEMENT OF

AGRICULTURE, | MANUFACTURES,
ARTS, | AND COMMERCE.

TO WHICH IS ADDED,

AN APPENDIX;

CONTAINING

A PROPOSAL FOR THE FURTHER IMPROVEMENT OF AGRICULTURE,

BY A MEMBER OF THE SOCIETY.

AND

THE TRANSLATION OF A LETTER FROM DR. TISSOT TO MONS. HIRZEL, IN ANSWER TO MONS. LINGUET's TREATISE ON BREAD-CORN AND BREAD.

BY ANOTHER MEMBER OF THE SOCIETY.

VOL. I.

THE FOURTH EDITION.

BATH, PRINTED, BY ORDER OF THE SOCIETY,
BY R. CRUTTWELL;
AND SOLD BY C. DILLY, POULTRY, LONDON,
AND BY THE BOOKSELLERS OF BATH, BRISTOL, SALISBURY,
GLOCESTER, EXETER, &c. &c.

M DCC XCII.

PREFACE.

THAT the publick may with greater clearness apprehend the scope of the present work, it is thought necessary to prefix the following short account of the nature and the occasion of its publication.

In the Autumn season of the year 1777, several Gentlemen met at the City of BATH, and formed a Society for the Encouragement of Agriculture, Arts, Manufactures, and Commerce, in the Counties of SOMERSET, WILTS, GLOCESTER, and DORSET, and in the City and County of BRISTOL.

This scheme received immediate approbation and great encouragement, not only by liberal subscriptions, but also by many useful communications of knowledge, both scientifick and practical, from ingenious and sensible correspondents.

On application to the London and Provincial Societies in this Kingdom, inſtituted for the like purpoſes, they very politely offered their aſſiſtance in communicating what might be generally uſeful; and to ſome of them we are indebted for much intereſting intelligence.

As the diffuſion of uſeful information in general is one end propoſed by this inſtitution, the Society think they cannot fulfil this intention in a more effectual manner, than by the publication of ſuch papers as appear to contain what is moſt likely to be of publick utility. Indeed, this is the only method by which the various improvements, and practical information, ſuggeſted to them, can be generally diſperſed, even among thoſe whom, from the nature of their inſtitution, they are under particular obligations to ſerve.

In ſelecting the following papers, regard has been principally had to ſuch as relate to matters of practice. Useful hints, however,

of

of the speculative kind, which may in their consequences lead to practical improvements, have not been neglected;—such will always be esteemed valuable communications, 'tho' inferior to those that have already been submitted to the test of experiment.

In a work of this kind, to be explicit and intelligible, are all the requisites with respect to language; and therefore the thoughts of our correspondents are generally given in their own words.

The Society, however, think it necessary here to observe, that although they have no cause to distrust the knowledge or veracity of any person who has favoured them with his correspondence, yet, for obvious reasons, they do not mean as a body to vouch for the truth of any relation, or to give authority to any opinion contained in the following papers, further than the *notes* express, and to recommend them as subjects of enquiry and examination.

To

To many ingenious correspondents, the Society take this opportunity of acknowledging their obligations, and of respectfully requesting such further information, both from them and others, as relates to any of the interesting subjects to which their attention is directed.

CONTENTS.

CONTENTS.

	PAGE
ON *setting* Wheat in Norfolk	1
On the same	5
Answers to the Society's Queries on setting Wheat	10
On *setting* Wheat as practised in Norfolk and Suffolk	13
Brief Account of the Norfolk Husbandry	19
On the Culture of Potatoes	26
On the same	30
On the same	32
State of Agriculture in the Isle of Wight	35
On the Disease called the Goggles in Sheep	42
Description of Mr. Boswell's newly-invented Machines for raking Corn-stubbles	44
On the Cultivation of Clover	49
The Society's Queries, with Answers thereto from the Sheriff of the County of Suffolk	51
On the Effect of Marl in Norfolk	62
On feeding Wheat in the Spring with Sheep	65
Method of making Reservoirs in dry Countries, for watering Sheep and Cattle	68
Experiments on Plants eaten or rejected by Cattle, Sheep and Hogs	70
On the Bulk and Increase in Growth of some remarkable Timber Trees	74
Mode of Farming pursued by a Member of the Bath Society	82

On the best method of raising Elms for Fences; manuring Fallows for Wheat; and preventing the Ravages of the Fly on Young Turnips — 90
On a peculiar Species of Grass found in Wiltshire 93
Observations on Thistles — — 96
On a Disease the Stock Lambs in Norfolk are liable to from eating *self-sown* Barley in Autumn — 103
Observations on the Mnyum Moss — 104
On the superior Quality of Grain produced from Set Wheat, to that sown Broad-cast — 105
Account of the Cultivation of Siberian Barley 106
On the Effect of Fern Ashes as a Manure for Wheat 113
On the Cultivation of Heathy Ground — 116
Instructions for the Prevention and Cure of the Epizooty, or contagious Distemper among Horned Cattle — — 120
On the Construction and Use of Machines for floating Pastures, and draining Wet Lands — 126
Use of Soaper's Ashes and Feathers as Manures 129
On planting Boggy Soils with Ash; and the Slopes of Hills with Forest Trees — — 131
Mode of cultivating Turnips in Suffolk — 133
On raising Potatoes from Seed — 136
On the Mode and Advantages of extracting the Essence of Oak Bark, for tanning Leather — 139
On drilling Pease — — 144
On the Culture of Siberian Barley — 146
On a new Oil Manure — — 149
Mode of weaning and rearing Calves — 153
On raising a Crop of White Oats and Grass Seed 154
Answers to the Society's printed Queries, from Gloucestershire — — — 156
On the great Increase of Milk, from feeding Cows with Sainfoin — — — — 163

On the Succefs attending the Planting Moor Land
  with Afh Trees — — — 165
Ufe of Stagnant Water as Manure — 168
Of the Management of Clover in Suffolk — 171
Thoughts on the Rot in Sheep — — 175
On the Mode of cultivating and curing the *Rheum-
  Palmatum*, or true Rhubarb — — 183
On the Cultivation of Rhubarb — 188
On the Cultivation and Cure of Rhubarb — 193
The fame continued — — — 196
Dr. Lettfom's Letter on Rhubarb — — 199
Dr. Hope's Letter on Rhubarb — — 200
On the Growth and Application of Rhubarb — 201
On the Extirpation of Plants noxious to Cattle on
  Dairy and Grazing Farms, &c. with Hints on
  the breeding and rearing Milch Cows — 206
On the Culture of Carrots, with Thoughts on
  Burnbaiting on Mendip Hills — — 214
Dr. Falconer's Report to the Society, on exami-
  ning fome of the Rhubarb cultivated in Somer-
  fetfhire — — — 220
On the beft Mode of deftroying Vermin, and pre-
  venting the deftruction of young Turnips by
  the Fly — — — 223
On the Culture of Carrots, and the Rot in Sheep 231
An Abridgement of feveral Letters publifhed by
  the Agriculture Society at Manchefter, in con-
  fequence of a Premium offered for difcovering
  the Caufe of the Curled Difeafe in Potatoes - 236
Defcription of, and Obfervations upon the Cock-
  Chaffer, both in its Grub and Beetle State — 258

APPENDIX.

APPENDIX.

A Propofal for the further Improvement of Agriculture; by a Member of the Society — 271

A Tranflation of Dr. Tiffot's Letter to Monfieur Hirzel; in Anfwer to Monf. Linguet's Treatife on Bread-Corn and Bread; by another Member of the Society — — — 317

LETTERS

LETTERS

TO THE

BATH AND WEST OF ENGLAND AGRICULTURE SOCIETY.

ARTICLE I.

On the Rise, Progress, and Mode of SETTING *Wheat in Norfolk.*

[By a Gentleman near Norwich.]

GENTLEMEN,

IT is with much pleasure that I now answer your enquiries relative to the practice of Setting Wheat in this county. It is, in my opinion, one of the greatest improvements in Husbandry that hath taken place this century; and, were it generally adopted throughout the kingdom, would be attended with very great advantages to the publick.

The Setting of Wheat seems to have been first suggested by planting grains in a garden from mere curiosity, by persons who had no thought or opportunity

opportunity of extending it to a lucrative purpose; and I do not remember its being attempted on a larger scale, till a little farmer near Norwich began it about twelve years since, upon less than an acre of land. For two or three years only, a few followed his example; and these were generally the butt of their neighbours' merriment for adopting so singular a practice. They had, however, considerably better corn and larger crops than their neighbours: this, together with the saving in seed, engaged more to follow them; while some ingenious persons, observing its great advantage, recommended and published its utility in the Norwich papers. These recommendations had their effect. The curiosity and enquiry of the Norfolk farmers (particularly round Norwich) were excited, and they found sufficient reason to make general experiments. Among the rest was one of the largest occupiers of lands in this county, who set fifty-seven acres in one year. His success, from the visible superiority of his crop both in quantity and quality, was so great, that the following autumn he set three hundred acres, and has continued the practice ever since. This noble experiment established the practice, and was the means of introducing it generally among the intelligent farmers in a very large district of land; there being few who

now

now *sow* any Wheat, if they can procure hands to *set it*. It has been generally obferved, that although the *set* crops appear very thin during the autumn and winter, the plants tiller and fpread prodigioufly in the fpring. The ears are indifputably larger, without any dwarfifh or fmall corn; the grain is of a larger bulk, and fpecifically heavier per bufhel, than when fown.

The lands on which this method is particularly profperous are, either after a clover ftubble, or on which trefoil and grafs-feed were fown the fpring before the laft, and on which cattle have been from time to time paftured during the fummer.*

Thefe grounds, after the ufual manuring, are once turned over by the plough in an extended flag, or turf, at ten inches wide; along which a man, who is called a *dibbler*, with two fetting-irons, fomewhat bigger than ram-rods, but confiderably bigger at the lower end, and pointed at the extremity, fteps backwards along the turf, and makes the holes about four inches afunder every way, and an inch deep. Into thefe holes the droppers (women, boys, and girls) drop two grains, which is quite fufficient. After this, a gate, bufhed with thorns,

* We cannot approve the cuftom of feeding clover previous to its being planted with wheat, as preferable to mowing the grafs.

is

[4]

is drawn by one horse over the land, and closes up the holes. By this mode, three pecks of grain is sufficient for an acre; and being immediately buried, it is equally removed from vermin, or the power of frost. The regularity of its rising gives the best opportunity of keeping it clear from weeds, by weeding or hand-hoeing.

In a word, this practice is replete with greater utility than any that has been made in the agricultural art.

In a parochial view, it merits the highest attention, as it tends greatly to lessen the rates, by employing the aged and children, at a season too when they have little else to do. It saves to the Farmer, and to the public, six pecks of Seed Wheat in every acre, which, if nationally adopted, (without considering the superior produce) would afford bread for more than half a million of people.*

The expence of setting by hand is now reduced to about six shillings an acre, and a very complete Drill-Plough has lately been introduced among

* This is a consideration of the utmost consequence, especially when Wheat is dear. We are sensible of the utility of this method, and warmly ecommend its being generally adopted in the Western Counties.

us, and found to answer extremely well, on summer fallows, by which the difficulty of getting hands is obviated, and the expence lessened, as with this plough one man can set an acre per day. The maker is Mr. James Blancher, of Attleborough in Norfolk.†

    I am, &c.

Article II.

On Setting Wheat.

[By a Gentleman Farmer in Norfolk.]

Gentlemen,

THE practice of setting Wheat in the eastern part of Norfolk is pretty general. The skill and labour required in performing it are so little, that it is done in many places by women and children only; in consequence of which there are few places so thinly inhabited, but the Farmer may obtain hands sufficient to do it upon the largest

---

† There is one of these Drill-Ploughs at the Society's-Rooms, with some improvements made by the inventor since the above letter was written. It has been tried by the Agricultural Committee in a field, and found to deliver the grain with great exactness and regularity, to the satisfaction of the Gentlemen and Farmers who attended the Experiment. Any person disposed to have one, may be furnished with it, by applying to the Secretary, price five guineas and a half, and carriage.

scale; and the expence is now reduced to eight shillings per acre on the average.* Each dibbler, employing three droppers, will set half an acre a day, making eight holes† in the length of every foot of the flag, whereby two dibblers with six droppers will find full employment for one plough, which, however, is not very material, as there will be no loss of time on that account, for the land may be all ploughed and set as soon as convenient. The advantages attending this practice are, the saving of a considerable quantity of seed, six pecks per acre at least—obtaining cleaner and better corn—providing a very lucrative employment for many of the poor, who would at that season have little to do—and gaining a greater produce. The last-mentioned benefit, however, I assert on no better ground than that of two experiments only, but these were conducted with so much care as to be nearly decisive.

These trials were made in the years 1774 and 1775, in the following manner:——

About Michaelmas 1774, a field of clover and ray-grass stubble, containing twelve acres, was

---

* It is now done for 6s. per acre very well, nearly an acre being done in a day in some places.   † Six are sufficient.

broken

broken up and ploughed into broad ftetches,* (the land being found and dry) which ftetches were alternately fet and fown throughout the whole field, and the corn after reaping was laid and carefully kept feparate. On threfhing, it was found that the Wheat which had been fet produced two bufhels per acre more than that fown.

About Michaelmas 1775, the like experiment was tried in a field of eight acres, which was a clean clover ftubble, treated in all refpects as the former. The refult was, that the produce of the Set Wheat exceeded that of the Sown Wheat one bufhel per acre.

In both trials, the corn of the Set Wheat exceeded that of the Sown in quality. It was more equal, and proved by far the beft; fo that, independent of the advantage accruing to the Farmer from fupporting the poor by *employment*, (the beft mode of fupporting them) inftead of affifting them from the parifh fund, (perhaps the worft) you will fee, as I fhall now ftate the account, that there is fufficient inducement from the immediate profit for him to attempt and perfift in this practice.

* Divifions by furrows.

	£.	s.	d.
Profit by feed faved on fix acres, being nine bufhels, at 5s. per bufhel — —	2	5	0
Ditto by increafe of produce,—fay fix pecks per acre on an average,—nine bufhels -	2	5	0
	4	10	0
Expence of fetting fix acres, 8s. per acre -	2	8	0
	£.2	2	0

which is feven fhillings per acre. But there are other advantages that I have not yet mentioned. A very great one I experienced laft year, when, from the heavy rains which fell in the fummer, all my fown wheat was more or lefs laid, none of my fet wheat was at all fo; by which I fuffered lefs lofs in reaping the latter than the former, and the corn was much fuperior in quality.

It has been found, that on ray-grafs ftubbles, or lands foul with twitch-grafs or other weeds, the corn being fet on the *middle*,* inftead of running (as it does by fowing) into the fpace between the

* This is an excellent remark, and ftrictly juft, although we apprehend feldom attended to by thofe who, either from prejudice againft this improvement, or from never being eye-witneffes of the great advantages arifing from the practice, have haftily and inconfiderately cenfured it:——

Though we apprehend few good farmers would fow lands which are foul with couch or twitch; and as to the annual weeds, the produce of all lands, they can do no harm at wheat feed time.

edges

edges of the flag, comes up free from the impediments of grafs and other trumpery which ufually environ it in the other method; and to this circumftance it is perhaps owing that, in the two experiments I have related above, the produce on the clover and ray-grafs ftubble exceeded that of the clean clover ftubble one bufhel per acre.—Hence fetting Wheat feems peculiarly advantageous to grafsy and foul lands;—a lucky circumftance, as the number of flovenly Farmers fo greatly exceed that of the neat ones.

I hope what I have faid on this fubject will be fufficient to fet the matter in its true light. I am fure it appears fo to me; for as I wifh not to fupport fyftems, I have felt no prejudices in favour of any merely as fuch. You are, therefore, welcome to make any ufe you pleafe of this letter, concealing only my name, as from the hurry in which it is written, I am afraid it may contain many inaccuracies.

I am, &c.

NORFOLCIENSIS.

March 5th, 1778.

Article III.

Answer to Queries respecting Set Wheat.

[By the same.]

Gentlemen,

I AM much honoured by the approbation which the information contained in my last letter† met with from your Society; and in compliance with your wishes, shall most readily answer the queries sent me by your Secretary, relative to the experiments I therein mentioned.

Query 1st. What kind of soil was the Wheat set in, and what the annual value of the land?

Answ. The soil was light, inclining to sand—worth fifteen shillings an acre, being within five miles of Norwich.

Q. 2d. How long had the land been in clover and ray-grass before it was broken up and set with Wheat?

A. One year only: in this part of the country, we seldom suffer clover alone to remain longer;

† See preceding Letter.

the

the natural grafs after that time getting so much posseffion of the ground as to render the second year's crop of clover of little value.

Q. 3d. How deep were the holes dibbled, and at what distance were the rows from each other?

A. The holes were dibbled about an inch deep, and the rows were two on a flag, near four inches from each other.

Q. 4th. How many grains were dropped in a hole, and was the crop hoed?

A. Two grains were intended to be dropped, but this is often uncertain, from the unskilfulness or carelessness of the children who drop the corn. This crop was not hoed, which, although an excellent practice and much used here when wheat is sown broad-cast, does not appear so necessary when it is set.

Q. 5th. How many bushels per acre was the produce of the two fields?

A. The experiment having been made with a view only to ascertain the proportional produce of the two methods, although that result was registered, I find, on a fresh enquiry, that the total produce

duce is forgotten. The common average growth of Wheat on the farm was about twenty-four bushels per acre.

Q. 6th. Was the straw of the set wheat higher and stouter than that of the sown?

A. The straw of my set wheat has always been stronger and higher; and being clearer from weeds, and of more equal size and strength, is more easily reaped.

Q. 7th. Do the Norfolk millers prefer the set to the sown wheat; and is it more bulky in the kernel, or whiter in the flour, or both?

A. Those of whom I have asked the question prefer the set wheat to the sown. In general it is the most even sample, having less small corn intermixed with it, otherwise nearly the same size as the sown wheat. I have not heard it generally observed that the flour is whiter.

Q. 8th. On what kind of land does the setting of wheat answer best upon the whole?

A. This has not yet been fully ascertained: I am, however, inclined to think that the produce of the set wheat would be greater in proportion to that of of the sown, on all ray-grafs lands, or such as are foul with twitch-grafs or other weeds. I mean that the difference would be greater than on any other cleaner lands; and this seems confirmed by the result of the two experiments mentioned in my last letter. Where the land is so stiff and wet that it cannot be readily covered by the bushes drawn over it for that purpose, I apprehend the corn would be better sown broad-cast.

NORFOLCIENSIS.

March 29, 1778.

ARTICLE IV.

On SETTING *Wheat, as for some years past practised in Norfolk and Suffolk.*

[By a Gentleman Farmer in Suffolk.]

GENTLEMEN,

THE desire of being useful to society leads me to communicate to you the following account of a new practice in agriculture, which is become general in Norfolk, and gaining ground fast in this county.

In October, the lands which have produced broad clover or artificial grasses, and sometimes old

old pasture the foregoing summer, are ploughed up; taking care to lay the furrows as even as possible. A heavy roller is then passed over it; and a man, or several men, each with a pair of instruments called dibbles,* walk backward, making two rows of holes on the earth or flag turned out of each furrow, so that the holes are three inches distant in the rows, and the rows on each flag or line of turf near five inches from each other. One pair of dibbles employs four women or children, who follow the man, and drop two grains into each hole. After this, a hurdle, covered with bushes, is drawn by a horse across the field till the wheat is covered, and the holes are filled up. In this method, the seed is regularly placed in the ground, four pecks† being sufficient for an acre, whereas ten pecks are usually sown broad-cast.

An experiment was lately made in Norfolk, and the following particulars laid before their Agriculture Society:——

"A whole field was *sown* and *set*, in alternate stetches. The sown wheat was cut, carried, and threshed separate from that which was set. The

* Short sticks with handles like a spade, and pointed in the form of a sugar loaf, with a cross pin near the point, to prevent the holes being made too deep.

† Three pecks have been found quite sufficient.

produce

produce of the *set* part was eight* bushels per acre more than the *sown*; and declared to be sixpence per bushel better."

I myself have this year *set* twenty-three acres, nine of which are old grass-land, ploughed this season for the first time; seven acres are a lay of seven years; and the remaining seven have been in grass only two years. The whole work (viz. ploughing, rolling, setting, and harrowing) was performed in seventeen days, by three ploughs, having a pair of horses and one man to each plough, five men dibbling, and twenty children dropping; the roll and bush-harrow employed another man. The land was as follows: The first field a black moorish soil, with a clay under it—this, with us, is called a woodcock soil. The next seven acres (a hill) were on the top a strong clay, and the lower part a mixed soil. The last seven acres, a light rich land.

* The difference of eight bushels per acre is so great, that we were inclined to think there must have been some error in transcribing the experiment; and accordingly wrote to the Gentleman who favoured us with the account. He has since informed us, that on repeated enquiry of the parties who set, reaped and threshed the corn, he is assured it was matter of fact, and has not any reason to doubt the veracity of those who related it. It was, however, a singular instance, for which we cannot account, otherwise than by suggesting that the corn on the stetches sown broad-cast, being when sown left *uncovered*, might be a great part of it devoured by birds and vermin before it was harrowed in.

We

We plough very strong land with a pair of horses a-breast, and one man holds the plough, and guides the horses with rope-reins.

For one acre in the Norfolk experiment.

	£.	s.	d.
Seed saved 5 pecks, at 6s. per bushel —	0	7	6
Superior produce 8 bushels, 5s. per bushel -	2	0	0
Superior value to that sown 6d. per bushel -	0	4	0
	2	11	6
Expence of setting —	0	10	6
Balance in favour of setting £.2	1	0

So that the farmer by this method gains the above balance, benefits society nine bushels and one peck, and at the same time feeds 25 extra mouths.— What a satisfaction to the benevolent mind!

When wheat is very full of weeds, it is customary with neat farmers to hoe it by hand when either set or sown, and they gain more by it than the 7s. 6d. per acre paid for the labour.

Fallows, or summer-lays, (as we call them) on heavy land, are constantly either sown with turnips, or planted with cabbages. The turnips are drawn,

drawn, and the cabbages are cut and carried to an inclosure, either at the barn-door or at the corner of the field, made with bush faggots, and well littered with straw or haulm; and I am certain that ninety acres, one-third or perhaps half ploughed, will maintain, by means of turnips and artificial grasses, at least as many cattle as the whole would in grass.

Cabbages make bad-tasted butter, but are excellent for fattening cattle, having an astringent quality so opposite to that of the turnip, that six weeks are saved in the fattening a beast; in which instance not only the saving of time, but of feed also, is of no inconsiderable consequence to the farmer.

The sort principally raised is the tallow-loaf, or drum-head cabbage, but it being too tender to bear sharp frost, I planted some of this sort and the common purple cabbage used for pickling (it being the hardest I know of) alternately, and when the seed-pods were perfectly formed, I cut down the purple, and left the other for seed. This had the desired effect, and produced a mixt stock of a deep-green colour with purple veins, retaining the size of the drum-head, and acquiring the hardness of the purple.

We

We have tried the Scotch cabbages, but found them so slow in their growth, that they would not answer unless sown in autumn, and planted in beds, the expence of which would destroy the profit.*

It is to be remarked, that frosts and cold winds, in this eastern part of the kingdom, prevent our sowing till the end of February always, and often till sometime in March; and the sort I have mentioned does very well at that time, requiring only to be taken out of the seed-bed, and planted in the field at eighteen inches distance every way.†

Turnips are always twice, and often three times hand-hoed with a nine-inch hoe. This work requires considerable dexterity, as the plants that are not to be cut up should be left regularly at a foot distance from each other, and the hoeing must be begun before the plants are too forward. A farmer who is not used to this practice, and sees the wi-

---

* We beg leave here to observe, that sowing the seed on a wheat stubble ploughed up, would do no detriment; on the contrary, it would prepare the land for a spring turn and summer tilth for turnips; and whether the seed be sown spring or autumn, transplanting and hoeing will be equally necessary. Wheat sown in autumn is no more trouble than in spring. But we find that Scotch cabbages, sown in April, come to their highest perfection both in size and quality in the county of Somerset.

† Surely the plants must be hoed; and the same process only is required in the Scotch as other kinds.

thered

thered plants the day after hoeing, will be frightened, and think his whole crop is destroyed, as I have experienced, but without any reason; for by this process they will come to 12, and even 18 or 20 pounds weight without their leaves. But in order to have them of this weight, the manure must be different from any thing I have seen in the West of England; and perhaps to render turnips as valuable there as they are with us, agriculture must be brought to the same degree of perfection as it is in Norfolk, Hertfordshire, some parts of Essex, and this county. The price of hoeing is 4s. 6d. the first time, 2s. 6d. the second, and 2s. the third, when a third hoeing is necessary.

I am, &c.

Suffolk.

Article V.

A brief Account of the Norfolk Husbandry.

[By a Gentleman near Norwich.]

Gentlemen,

IN answer to your enquiries respecting the improvements in husbandry made in this county, be pleased to accept the following account.

About

About sixty years since, a great part of this county was sheep-walks, rented only at about eighteen-pence an acre; and even within my memory, many thousand acres were in this state, which now are turned into the finest farms, and let at twenty shillings per acre.

The late amazing improvements may be attributed to various causes. Among others, the following have not been the least operative.

1*st.* Inclosing our heath and waste lands; folding sheep; and the most extensive use of marle and clay, on sandy soils especially.

2*dly.* By the general introduction of turnips, well hand-hoed; of clover, ray-grass, and buck-wheat, and an excellent course of crops.

The farms being generally large, and held on long leases, the tenants were thereby enabled to lay out their money freely in improvements, without being in danger of losing the advantages arising from their cost and labour.

We possess one natural advantage, which, perhaps, cannot be found in an equal degree in many other counties. In all our sandy lands, wherever

we

we dig, we find excellent white and yellow marle, or clay. The goodnefs of the marle is determined by its fubfiding quick in water. On the firft difcovery of marle, our farmers fpread it in larger quantities than at prefent; few laid on lefs than eighty loads per acre; but for near thirty years paft, the general quantity has been from forty to fifty loads (or tons) per acre. The effects of this quantity will laft twenty years; and then half as much more added will reftore fertility to the foil. We have however found, that, on lands wholly fandy, clay has had a better effect than marle; but where the foil is a mixture of fand and loam, or of fand and gravel, marle does excellently. It is not, however, to marle and clay only, that our improvements are owing. Our fheep are folded both fummer and winter. We fatten beafts during the winter on turnips in our farm-yards, in which we alfo keep a large ftock of fwine. Our ftubbles are cut, and, with large quantities of ftraw, converted into manure. Oil cake is alfo laid on wheat lands to the amount of two guineas per acre.—Thefe manures, freely ufed, have proved the fources of wealth to thoufands.

The ufual courfe of crops among our greateft and beft farmers, is, 1. Turnips; 2. Barley; 3. Clover, or clover and ray-grafs; 4. Wheat. This courfe courſe has of late years become very general, and keeps the ſoil clean. We manure for turnips if poſſible, and alſo for wheat. Sometimes our clover is extended to three years, but not frequently. Of late, eſpecially, our clover often fails the third year, and ſometimes the ſecond, if the land be wet; for wherever the water ſtands in the winter or ſpring, clover turns black and decays. Our farmers agree in the opinion, that if turnips are ſown on a well-conditioned fallow, and twice hoed, and the land ploughed three times for barley, the clover may remain at leaſt two years without giving a foul crop of wheat, eſpecially as our wheats, on clover lays, are of late almoſt wholly *ſet*, and more eaſily kept clean than when ſown broad-caſt. We ſet from two to three pecks per acre, and find great advantage from the practice—the expence of ſetting by hand is, from ſix to eight ſhillings per acre. On our fallows, we plant with Mr. Blancher's drill-plough, at leſs than half the expence, and with equal regularity and ſucceſs.

The Norfolk huſbandry is, as Mr. YOUNG has juſtly obſerved, quite a ſyſtem, every ſucceſſive part of which is dependant on the foregoing, and therefore it will not admit of much variation.

As

As every thing depends on the succefs of turnips, *their* succefs depends on good hoeing. They are the only fallow in our ufual courfe: nor can we change them for a *mere* fallow, becaufe the sheep, kept to fold, and to feed off the clover and ray-grafs, would then ftarve. We give four ploughings for turnips, and hoe them well twice. They often, with this culture, prove worth five guineas an acre. The principal part of the crop is drawn and carried into farm-yards for fattening beafts; the remainder we feed off with sheep and lambs, which clear the land of every part of them.

We generally mow the firft and fecond growth of clover; not merely on account of the hay, but becaufe, by repeated experience, we are convinced the wheat which follows is far better than it would be after feeding.

Soaper's afhes are laid on ftrong wet lands with great fuccefs; and alfo on paftures as a top-dreffing in the beginning of April.——Malt-duft and foot are ufed on meadows, and anfwer well; the latter is purchafed at high prices from Norwich.

The winter food of cows is chiefly turnips and ftraw, in the farm-yards, which are kept well littered with chopt ftubble and ftraw.

We reckon six horses necessary for one hundred acres of arable; and with two in a plough we till two acres in a day; five or six inches deep. Stubbles for fallow are ploughed in during autumn—this also destroys the weeds.

A good dairy-maid with us will take proper care of twenty cows; and to every cow our best farmers keep one hog.

The common mode of estimating the expence of taking a farm is, that three rents will about stock it, or four very completely.

In some parts of this county considerable quantities of cole-seed are raised; we hand-hoe it like turnips, and by that means nearly double the value of the crop.*

Our broad clover sometimes produces near three tons the first cutting per acre. Nonsuch, ray-grass, and small white clover, are an excellent mixture to lay down dry lands with; and yield the sweetest hay.

* An Essex Gentleman informs us, that he sows cole-seed in drills, one foot or fourteen inches apart; and that by this method the largest quantity and the best seed are produced.

Near

Near the coast great quantities of sea-weed, or ooze, are collected, and used as manure to good purpose. We mix it in compost with earth and lime, or marle and dung, for one year, and then lay it on arable land. Our best farmers beat down thistles and nettles, and mow the weeds in their borders, ditches, and the adjoining roads, lanes, and commons, before they seed, and burn them to ashes; the ashes are used as a top-dressing for their meadow-lands. This is excellent management, and worthy of general imitation; for it saves infinite labour the succeeding spring in the fields adjoining.

Most of the farmers round Norwich carry dung to the distance of ten or twelve miles. They load a waggon for two shillings, or a cart with three horses for one shilling.

A great deal of buck-wheat is sown here as a preparation for wheat, and answers well. Six pecks are sown per acre, and the average produce is from three to four quarters. The price is generally the same as that of barley, and it is an excellent fattening for swine and poultry.

Many of our farmers have cultivated lucerne with success on good rich lands. On a poor soil it seldom answers well.

Two-

Two-wheeled ploughs are used in general, as being most easy and expeditious; but in heavy lands they use swing-ploughs, and two horses always do the work. We should smile at the folly of putting four horses to a plough in *any soil,* because we know it to be unnecessary, except where the land abounds with stone.

I am, &c.

[The preceding Letter abounds with much useful information, and the writer seems thoroughly to understand his subject.]

Article VI.

On the Culture of POTATOES.

[By the Rev. Mr. HIGSON, Vicar of Bath-Easton.]

GENTLEMEN,

HAVING had many years' experience in cultivating Potatoes, I take the liberty of sending a few observations thereon, which are much at your service.

Potatoes flourish most in a dry sandy loam. The ground should be well manured with rotten horse-dung; and the seed changed every year.

The Cheshire or Lancashire kinds answer best in the neighbourhood of Bath. Warminster or Farringdon potatoes are also a good change for this soil. Those from Monmouth and the adjacent parts are not so good.

They should be planted in fresh ground every year. If either fresh or the same seed be planted upon the same soil for two or three years successively, the crops will generally fail, the haulm come up curled and blighted, and the roots will be worm-eaten and cankered. The cause of this may perhaps be assigned. Every species of plant is provided by nature with pores of such construction and magnitude, as are capable of receiving those particles of nourishment only whose dimensions are correspondent to the said pores: Hence every species must receive or imbibe the abovesaid particles only, and reject all others; and, consequently, if the same species be planted, or sown, upon the same soil, for two or three years in succession, the greater part of such particles will be exhausted, and the plants cannot flourish for want of proper nourishment.

Potatoes should not be planted deeper than four inches or four inches and a half, and the seed or sets should lie one inch or one inch and a half above

above the dung. Whole potatoes fhould be planted at the diftance of two feet and a half or three feet fquare: cuts at the diftance of eighteen inches fquare.

I have feen potatoes planted in this parifh upon ground without dung ten or twelve inches deep, and at the diftance of eight or nine inches. Thefe crops have always failed, and, as I fuppofe, for want of proper nourifhment.

I have tried the following experiment for thefe five years laft paft:—The firft year, on the fame day, and in the fame ground, I planted whole potatoes in ranks, at the diftance of three feet fquare, and cuts of the fame kind at eighteen inches fquare. For the laft four years I have planted whole potatoes at the diftance of two feet and a half fquare, and cuts at eighteen inches. The whole fets were earthed up three or four times; (*i. e.* as long as the haulm would ftand) and a few ranks of the cuts were earthed up alfo. The whole fets have always produced a greater crop than the cuts, in proportion to the quantity of ground; and the potatoes have been larger and fairer. I have obferved little or no difference in the produce of the cuts, whether the ranks be earthed up or not. This, I think, may be thus accounted for:—if any benefit is to be

received

received from earthing up a plant, it must be because more nutriment is thereby added through the pores of the haulm or stalk. Now there was sufficient room in the ranks that were planted with *whole sets* to earth them up equally on every side; but not so in the cuts, for the earth which was added on one side of the plant was taken away on the other.

There is a small white early sort of potatoe, which, of late years, has been much cultivated at Altringham in Cheshire. They plant this species in January, or as soon as the earth is dry and the weather mild. It never blows; but is fit for use a month or six weeks sooner than any other kind.

I have known the following experiment tried with good success:—They plant in October, and if there come any severe frost without snow, they cover the potatoes with pease-haulm, bean-stalks, straw, or other light covering. The whole crop is dug up in May, and another sort immediately put in, which is also dug up in October following. I have eaten new potatoes thus raised in April. This species hath been of late introduced into this parish from Altringham, but they plant too late, never cover from frost, nor change the seed.

I am, &c.   J. HIGSON.

ARTICLE

Artcle VII.
On the Culture of Potatoes.

[Communicated by the Norfolk Agriculture Society.]

Norwich, Jan. 13, 1776.

AT a general quarterly meeting of the Norfolk Society for the encouragement of Agriculture, B. G. DILLINGHAM, efq; in the chair;

A premium of three guineas was adjudged to Mr. *Joseph Wright*, of Great-Melton, for planting and gathering the beft crop of potatoes, the quantity of land being one rood, and the produce ninety-one bufhels.

The Society cannot but recommend, in the ftrongeft terms, a more general attention to the culture of this moft valuable root. In the prefent inftance, and that not a very extraordinary one, the labour and expence of the hufbandman are amply rewarded by the produce, as appears from the following particulars of Mr. *Wright*'s experiment. The public too muft reap advantages of the moft important nature, as the potatoe, befides being an excellent wholfome food in various modes of application, is well known to make a fweet and nourifhing bread when mixed in equal quantities with the flour of wheat.

It is almost unnecessary to observe to the intelligent farmer, that as good a crop of wheat may be expected after well-cultivated potatoes, as in any other mode of husbandry.

Calculation for cultivating one acre of land with potatoes, according to Mr. *Wright*'s expences, and valuation of this crop:——

	£.	s.	d.
One deep ploughing	0	4	0
Seed 32 bushels, at 1s. per bushel	1	12	0
*Manure 24 loads, at 2s. per load	2	8	0
Expence of setting, the weather being so dry that the sets were put in with an iron crow	1	0	0
Hoeing and earthing up several times	1	0	0
†Expence of taking up	5	8	0
Rent and other charges	1	10	0
	13	2	0
Produce 364 bushels, at 1s. per bushel	18	4	0
Clear profit	£.5	2	0

ARTICLE

* It should be observed, that only half the expence of manure ought properly to be charged to the potatoes, as the land remains in fine order for any succeeding crop. Nothing is charged by Mr. Wright for carriage of the crop to market; but as he has valued them at only one shilling a bushel, which is greatly under the usual market price, it may be supposed they were either sold at home, or that the expence of carriage, sale, &c. were deducted in that estimate.

† The price charged in this calculation for taking up the crop appearing to us extraordinary, our Secretary was directed to enquire
parti-

Article VIII.

Abstract of a Letter on the Culture of Potatoes.

GENTLEMEN,

OF all the roots produced in our fields and gardens, none appears to be of so much consequence as the potatoe. As food for the poor, they are certainly to be preferred to turnips; and might be rendered equally useful for cattle. As a fallow crop, they tend greatly to meliorate the soil;

particularly into that article. He did so; and received for answer, that Mr. Wright had all his ground ridged up when the crop was gathered, for the greater conveniency of getting them out as clean as possible;—that he also had the crop picked over and separated into two or three different sorts, which took up a considerable time;—and that he gave the people employed victuals and drink all the time; which were all included in the general expence mentioned.

From this explanation it appears, that an acre of potatoes, producing 364 bushels, may probably be taken up in the usual way for about half the expence he has rated it; and when half the expence of manure is taken also from his estimate, we think the nett profit of an acre so cultivated, and producing such a crop, would be at least nine pounds. It may perhaps be said, that in many places the rent of such land as is proper for this purpose would be higher than he has stated it; but, admitting this to be the case, the price of the crop when sold would exceed his estimate so much as to balance the account.

To these certain advantages, arising from the cultivation of potatoes as a food for our tables, we wish to call the attention of our readers to a still further use, in which they would be a very profitable production; we mean, for feeding swine and cattle. For these purposes, they are an excellent hearty food, and it has been proved by experience that swine and cattle will eat them very freely.

being in this respect equal to turnips; and, in general, pay the owner of the land much better.

Some persons have objected to the general cultivation of potatoes, from the apprehension of wanting a market; but while they are retailed out at two shillings and six-pence, or even at two shillings a bushel, (and it is very rarely that we can purchase them lower) this apprehension will be groundless.

Add to this, that potatoes may be very profitably used as food for cattle and hogs. No food is better for rearing and fattening the latter. Cows and oxen will also eat them freely, and they are more easily preserved from frost than turnips: Hence they would prove an excellent succedaneum at the season when spring food is most wanted.

If potatoes were introduced regularly in the farmer's course of crops, on light good soils, great advantages would ensue. He need not be at the trouble and expence of having them dug up clean. Let him only take up the best part, and then turn his swine in: they will gather and fatten on the rest, and repay their value in the manure they leave behind them.

Potatoes

Potatoes grow best in a soil that is loose and deep, where the swelling of the roots meet the least obstruction, and where they draw the greatest nourishment most easily. On this account, where the quantity intended to be raised is small, digging is preferable to ploughing. But if the land be ploughed deep, and well pulverized, success need not be doubted. They ought to be planted in lines, eighteen inches apart, and at twelve or fourteen inches distance in each line or row. This will give opportunity for earthing them up with the horse-hoe while young, which will greatly promote their fertility. If the horse-hoe is not intended to be used, plant them a foot square, and earth them up with hand-hoes several times, which, although more expensive, will repay the cost.

Care should, however, be taken, in the latter hoeings especially, not to go too near the plants, lest you cut the roots. You need only, after they are weeded by hand, to draw up the earth from the centre round each plant. Vast quantities may be obtained by a little additional labour in keeping them clean, and the land will be left in excellent order for succeeding crops. It is necessary to observe, that the same kinds ought never to be planted twice together on the same land; nor the

same

same land set with potatoes more than two years at the longest. When raised from the seed, great varieties, and some excellent new kinds, will be obtained.

I am, your's, &c.
T. PAVIER.

Monckton, near Taunton,
  March 1, 1779.

Article IX.

State of Agriculture in the Isle of Wight.

[Communicated by a Gentleman who lately visited that Island.]

Gentlemen,

HAVING lately been in the Isle of Wight, principally with a view to obtain some authentic information relative to the mode of husbandry and produce of that island, I have selected the following remarks from notes taken on the spot, and beg leave to present them to your Society.—I shall first notice

The Nature of its Soil.

This island is finely diversified with gently rising hills, and fertile well-watered valleys. The hills, being

being generally cultivated, and abounding with springs, are very fruitful; some few excepted, which afford good feed for sheep.

In some parts, the soil is gravelly, and abounds with flints like that of Hampshire. In some other parts, it is a strong clay; but in the general, it consists of a fine hazel loam, mixed with just enough sand to make it work kindly, and render it easily penetrable to the tender fibres of grain and plants.

Being surrounded by the sea, the air, and consequently the dews and vapours when condensed and falling in rain, are impregnated with salts, which add greatly to the fertility of the soil.

Rents are in general from sixteen to twenty-one shillings per acre; some meadow lands, near the principal towns, considerably higher.

Manures.

I was informed by several farmers, that, 'till within a few years past, lime was almost the only manure used in the island, farm-yard dung excepted: but of late chalk has been much substituted in its stead. The farmers find, from experience, that

chalk

chalk is more operative and lasting. Thirty waggon-loads, of forty bushels each, are generally laid on an acre, and last from ten to fifteen years. It is, however, the general opinion, that a second chalking is of little service. Their chalk is of a hard kind, both blue and white; the former, I think, might with greater propriety be called marle, as in texture and quality it very much resembles the blue marle found in Somersetshire; but with this difference, that in putting it in vinegar the effervescence is not so strong. Of this kind they lay twenty loads on an acre; and I was told that in lands thus chalked more than thirty years since, the benefit is still very apparent.

In sundry articles of good husbandry the farmers in general seem still very remiss. They neither gather in their stubbles, nor confine their cattle to the farm-yard in winter. Hence their stock of farm-yard manure is small in comparison of what it otherwise might be: add to this, that they are not in the practice of digging up the borders of their fields, or mixing up heaps of compost.

In those parts contiguous to Newport and West-Cowes, the soil is naturally the least fertile; but as great quantities of stable dung are made in those towns,

towns, and eagerly bought by the adjacent farmers, the deficiency is pretty well supplied, and the produce of their lands nearly equal to that in other parts of the island.

Their general inattention to sea-weed* is another proof of their deficiency in the knowledge of manures.

On the east, south, and west coasts, the sea beats with great violence, and throws up vast quantities of weeds on the beach, which might be collected with little expence, and being mixed in compost with lime and earth, or dung, prove a most valuable and fertilizing manure in those places where it is most wanted.

I enquired of divers farmers why they did not thus apply a treasure which nature had so amply furnished them with. The substance of their answer was, " that it never had been used unless to " mix with chalk for their bean lands, and that " they apprehended it promoted the growth of

---

* We cannot help expressing our surprise, that this excellent manure should be so much neglected as it is on the banks of the Severn; but it seems almost impossible to make common farmers sensible of the advantages arising from improvements, not made by their forefathers.

" weeds ;"

"weeds;" not considering that whatever promotes the growth of weeds must, for the same reason, promote the growth of grain, and that its fertilizing the soil is the cause of both.

Course of Crops, *and Produce per Acre.*

In the Eastern part of the island, the usual course of crops is, on a summer fallow,

1 Wheat
2 Barley or Oats
3 Clover and Ray-grass, one year
4 Wheat, Barley, or Oats

About the centre of the island, the following course is mostly adopted:—

1 Turnips   4 Wheat   6 Clover
2 Barley   5 Barley   7 Wheat.
3 Clover

On the North-West part, after a fallow,

1 Wheat
2 Barley, or Oats
3 Clover, and Ray-grass, or hop-clover two years:—

Or, which is a still better course, and similar to the Norfolk,

1 Turnips   3 Clover and Ray-grass
2 Barley   4 Wheat.

When they break up clover-lays for wheat, they seldom plough more than once; but give four ploughings to their fallows. They sow all broadcast, and in general reap from three to four quarters per acre.

For Barley, they plough three times, sow four bushels, and the produce is from five to six quarters per acre on an average.

For Oats, they plough but once, sow four bushels and a half, and the produce is about five quarters on an average in return.

But on the Southern part, particularly near Godshill, they sow oats after turnips, and reap from eight to nine quarters per acre.——These are very great crops, but the land in this part of the isle is a fine rich loam, which cannot fail of producing large crops of whatever grain is sown upon it.

On some of the stiff clays, they plant Beans; and, very unnecessarily, set nine or ten pecks per acre;

acre; for the setting of which they pay two shillings and four-pence per bushel. They never hoe them; but, notwithstanding this bad management, these clay-lands are so rich and well adapted to this species of pulse, that they produce in general five quarters per acre.

The *Turnip* Husbandry seems to be but imperfectly understood. In the northern part of the isle they are not hoed at all. In the eastern part, they plough four times, harrow, and hoe once. They feed them all off with sheep, and value the crop at fifty shillings per acre.——Strange it is, that the proper mode of culture, and real value, of this excellent root should be so little known or attended to!

Of *Clover* they generally cut two tons per acre, and then let it run to seed. Sometimes they sow tares or vetches after clover, and mostly cut them green for horses. In the south-west parts, they sometimes sow them for feeding sheep.

On some of their most sandy lands, they sow buck-wheat, but the quantity is small, and it is only used for fattening swine.

The pasture-lands, especially in the valleys, are very rich, and produce excellent hay, which is stacked ftacked in the neateft and beft manner I ever faw. But as moft of the farmers keep dairies, the greateft part of the grafs is fed off. To each Cow they allot one acre and a half of grafs. Their cows are moftly of the Alderney breed, many of which yield eight pounds of butter per week. Their winter food is ftraw before calving, and then hay.

Where the lands are wet, they make excellent covered drains, with chalk and heath or ling. I examined fome, and never faw any executed in a better manner. Many quick-hedges have been raifed within thefe twenty years, and are in fine order; but in ditching the farmers appear to be very deficient.

I am, &c.

May 8, 1779.   E. R.

Article X.

On the Difeafe called the Goggles *in Sheep.*

[By a Gentleman in Wiltfhire.]

Gentlemen,

WITHIN thefe few years, we have had a a difeafe among the fheep, now generally known by the name of the *Goggles*; a difeafe which has

has destroyed some in every flock round this county, and made great havock in many.

The sheep most subject to it are *two teeth*. It is not infectious, but hereditary, and undoubtedly runs in the blood. It has no affinity with giddiness, for they do not run round. It most resembles the staggers in lambs, with this difference, that whereas staggery lambs shew weakness before, and fall forward, goggly sheep shew a weakness behind, and fall backward, when forced to run.

When first observed to be diseased, their ears drop, and they rub their tails much more than other sheep; they then discover the weakness above-mentioned, and grow poorer and weaker till they cannot drag their limbs behind them, and at length die.

I have examined a few, and found the viscera all sound. I have blooded one, and found no inflammatory crust. I can neither myself imagine, nor find one who can venture even to conjecture, the cause.* As it is a matter of consequence, perhaps, were you to make it the subject of the two

* It has been suggested to us, that the seat of this disease, most probably, is in the spinal marrow.

following premiums, it might be a means of stopping its progress: the first, to the surgeon who shall dissect the greatest number of goggly sheep, and give the most accurate description, with the best observations on the disease; and the second to the person who shall discover an effectual cure.

I am, Gentlemen, &c.

Article XI.

Description of a newly-invented Machine for Raking Summer-Corn Stubbles.

GENTLEMEN,

ABOUT three years since I found some difficulty in procuring hands to take up my Lent or Summer corn in the method usually practised in this county; that is, by forking the swarths into cocks, and raking the ground with hand-rakes by women. Men are generally employed in forking it. It therefore occurred to me that an instrument might be made to answer the purpose of raking it by hand. I knew the Norfolk method of doing it by drag-rakes, (as they are called) drawn by men; but the men were wanting elsewhere.

where. I had often feen a horfe-rake, made for gathering the *gramen canine*, or couch-grafs, together upon fallow lands, and knew a farmer who had ufed it for his own mown wheat ftubbles; but this rake being drawn from the end of the beam by the horfe, dragging the ends of the teeth upon the ground, collected fuch quantities of weeds, grafs, earth, and ftones with it, as nearly to render the corn of no value; befides, it could not be ufed for clofe-mown ftubbles at all.

Having for many years ufed the Norfolk ploughs here, I thought a rake might be fo conftructed as to go on the breaft-work of one of thefe ploughs in the fame manner as the plough itfelf is ufed.

I therefore had one made nine feet and a half long, and the teeth fix inches afunder. Upon applying it in the place of the plough on the breaftwork, I found it anfwered extremely well, except that when it met with any confiderable obftruction at one end, it drew the other end aflant. To remedy this inconvenience, I took away the pillar (the part of the breaft-work that the beam refts upon, and which is raifed higher, or let down lower, to fink or raife the plough) and had another made to extend about a foot or rather more beyond

yond the outsides of the standards, and from each end of the chain, made to let out or take up at pleasure, to each end of the pillar: this kept the rake even and steady. To my great satisfaction, I found it succeed even beyond my expectation; for by means of this breast-work, it could, like the Norfolk plough, be instantly set up or let down to the greatest degree of nicety; so that any stubble, whether cut high or low, whether very full of grass or clover, or quite clean, might be raked by it with equal facility; for the teeth being made very much curved, the lower part of the back of the teeth rests upon the ground, and the points stand out of it. The weight of the rake presses the teeth close to the ground, and the corn is gathered into the throat of the rake, without digging up the weeds or the soil. The teeth are made sufficiently strong to prevent their bending. I have found a rake of the length above-mentioned very manageable; whether it would do better if made longer must be left to future experiments. I was determined to this length by the breadth of our gate-ways, being just enough to admit it through them without taking it to pieces.

For persons who want to remove it to a distant part, two small wheels might be added, to put on occasionally

occasionally at the ends, to raise the teeth from the ground as it is drawn along the road.

I am sensible, that if a low wheel were fixed at each end, even when in its work, it would greatly lessen the friction, and the horse would draw it the easier; but it would render it more complex, and, perhaps, occasion it not to turn so easily at the ends of the land. I have, however, had it in idea, to fix some kind of standard on the head of the rake for a line, like the Norfolk plough-lines, to come back to, that the man might guide the horse himself, and save the expence of a boy to lead him; but to this there seem to arise some objections.

One horse, and a boy to lead him, with a man to clear the rake, will easily rake twelve acres of stubble in a day; and if two horses are taken into the field, to be used alternately, twenty acres might be raked in the same time; but this would be hard work for the man.

The manner of using it is as follows:—

The rake being put on the breast-work of the Norfolk plough, in the same manner as the ploughs are, the horse draws it with the same traces, &c.

(only

(only in the plough two horſes are uſed, and here but one) and being ſet into its work to a proper height, according as the ſtubble is long or ſhort, the boy leads the horſe acroſs the ridges, the corn being previouſly put in cocks by the forkers, the man follows the rake, and when it is filled, he ſpeaks to the boy, who ſtops the horſe, and puts him back a ſtep or two. This is done that the man, by drawing the rake back a little, may the more eaſily and ſpeedily free it from the corn; then lifting it up, and the horſe inſtantly going on, he drops the rake juſt beyond the ridge thus gathered together. This he repeats as often as it is full, till he reaches the end of the land. Then he turns, and coming back by the ſide of the part raked, empties the rake adjoining to the other.

By this means the raked corn lies in ſtrait rows acroſs the field, and, when dry, is turned if neceſſary, gathered up, and carried away.

I am ſatisfied it might alſo be advantageouſly employed in raking upland hay-ground, and all ſorts of ſeed clover land.

It may not be amiſs to mention, that in the firſt rake I made, the teeth were only three inches aſunder.

asunder. I soon found they were too close. Taking, therefore, every other one out, I made another six inches asunder—the holes in the first not being filled up, the teeth might, if necessary, be replaced, and then would be thick enough to rake any gentleman's lawn which is kept frequently mown.

I am,

Your obedient servant,

GEO. BOSWELL.

Piddletown, Dorset.

[Mr. Boswell is Author of an ingenious Treatise, intitled, *An Attempt to reduce the modern Method of Watering Meadows into a regular System*.

ARTICLE XII.

On the Cultivation of CLOVER.

GENTLEMEN, *Dec.* 20, 1778.

AS I apprehend every experiment in Agriculture, which is attended with any remarkable degree of success, may be of some use when made publick, I take the liberty of transmitting the following to your Society.

A neighbour

A neighbour of mine, who is a very good farmer, had a field containing thirty-eight acres, the foil a cold wet clay, which, for some years after he held it, scarcely paid its rent. Determined, however, to try what could be done with it, he under-drained it, and, in the spring 1775, mended it with turf-earth, digged from the borders of fields and highways, mixed with stable-dung. In March he gave it a good ploughing, and sowed it with Zealand barley: after the barley came up, he threw in ten pounds of the common red clover per acre.* The advantage of the under-draining and manure were soon apparent. The barley was an exceeding fine crop, producing seven quarters per acre on an average throughout the field.

The following spring the clover shot early, and in the summer proved a very strong crop. In May he turned in all his cattle, which by the 10th of June, had fed it off quite bare. He then took them out, and let the clover stand for seed. The summer proving wet, it succeeded well, and the average produce of the field was seven bushels and a half per acre, the whole of which he sold at thirty-nine shillings per bushel—amounting to 555l. 15s.

* Six or seven pounds per acre is supposed to be a sufficient quantity.

As soon as the seed was off, he ploughed the field for wheat, and sowed it broadcast, with the red Lammas kind from Kent. The crop was excellent, and the produce four quarters per acre.

With the same husbandry, I had last summer thirty-nine bushels of clover-seed from three acres of land, notwithstanding the dryness of the season, but the land was somewhat better in quality.

J. B.

S———d, *Essex*, 1779.

Article XIII.

Circular List of Queries sent by the Society at Bath to the High Sheriffs of the different Counties in England; with the Answers transmitted by the High-Sheriff of Suffolk.

IN June 1778, the society formed the following list of queries relative to Agriculture; and directed them to be transmitted to the High-Sheriff of every county, requesting him to procure answers thereto from some of the best farmers, and send to the Society.

Queries

Queries from the Bath Agriculture Society.

1st. What are the kinds of soil from which you generally obtain the best crops of wheat, barley, pease, oats, beans, vetches, turnips, carrots, and cabbages? And what are the usual quantities of seed sown, and the average annual produce per acre Winchester measure?

2dly. What is the usual course of crops adopted by your best farmers on the different soils.

3dly. What manure now generally in use do you find most serviceable on the following soils respectively, viz. Stiff clays, light sand, gravelly, moory, cold and wet, or what is called stone-brash land? in what quantities are the several manures laid on per acre,—at what season,—and how long will each last without renewal?

4thly. Have you discovered any *new* manure more efficacious than those generally used, and which may be easily obtained in large quantities? if so, what is it, when and how applied?

5thly. What is the best top-dressing for cold wet pastures, which cannot be easily drained?

6thly.

6*thly.* What materials do you find beft and moft lafting for drains, or land ditches?

7*thly.* What are the kinds of wood which you have found from experience to thrive beft on bleak barren foils, cold fwampy bogs, and black moory ground?

8*thly.* What are your methods of raifing lucerne, fainfoin, and burnet?—On what lands do you find them to anfwer beft, and what the average produce?

9*thly.* How is your turnip hufbandry conducted; and what is the beft method of preventing or ftopping the ravages of the fly on the young plants?

10*thly.* Do you prefer the drill to the broadcaft method of fowing grain;—in what inftances, and on what foils?

11*thly.* What is the comparative advantage of ufing oxen inftead of horfes in hufbandry?

12*thly.* What have you found to be the moft effectual preventative or remedy for the rot in fheep?

1 3*thly.* What new improvements have you made or adopted in implements of husbandry?

To the above Queries the Sheriff of the county of Suffolk favoured the Society with the following Answers; which he informed them were given him by a very good farmer, and approved by all who had seen them.

To the first and second.—Good strong mixed soil. Wheat on clover-lay, after one year, once ploughed, and sown broadcast, with ten pecks per acre, well harrowed in;—average produce, from three quarters and a half to four quarters per acre.

To prepare for TURNIPS, the year following, summer-till the land; turn in the wheat-stubble about December a moderate depth, and let it rest till the March following; harrow it well; then turn it in somewhat deeper, below the first ploughing, the deeper the better; for turnips thrive best where there is a plenty of deep mould. In May repeat the harrowing, and turn it up with a fine rift balk. After it has taken the benefit of the sun, harrow it down, and gather out the spire-grass, &c. which should be burnt in heaps upon the land. If it be not clean, repeat this a second time; then give it a clean earth, and harrow it down. Manure it with

with twelve loads of short dung, or eighteen of long dung per acre. At Midsummer plough the dung in a good depth with a close furrow, and sow the seed close after the plough. Sow from one pint and a half to two pints and a half per acre, as the season and quality of the land may require. In a month the plants will be fit to hoe. When they nearly cover the land, hoe them a second time, with a seven or nine-inch hoe, and leave the plants at least fourteen inches asunder. The price of hoeing here is generally 4s. 4d. per acre the first time, and 2s. 2d. the second. Set them out at least a foot distant from each other. A good crop will produce from thirty to forty cart-loads per acre, which for many years past have sold for from three to four pounds per acre.

2*dly*. To prepare for BARLEY to lay in with Clover, plough the said lands in February, as they are preparing and clearing off the turnips. Two *stirring* and one *sowing-earths* will be sufficient. Three bushels per acre, well harrowed in, will be a good feeding. Then throw in broadcast from nine to twelve pounds of clover-seed* per acre,

---

* We apprehend six or eight pounds of clover-seed would be fully sufficient; and that the clover should not be sown earlier than a fortnight after the barley. If they are sown together, the clover, in rich lands especially, will be apt to get above, and choak the barley crop.

struck over with a light harrow. Roll it down, or otherwife, as the feafon proves wet or dry. Average produce, from three to five quarters per acre. The following year clover, two crops in the feafon; firft mowing in June, the latter in September; the general produce from three to four tons per acre. In October fow the clover-ftubble with wheat, as above directed, without manuring, or it will be winter-proud if the land be rich.

3dly. To improve ftiff Clay Lands, lay on coarfe wafh-fand, cinder-duft, wood-afhes, ftreet-dirt, or ant-hills quartered, taken up and burnt. Thefe mixed together, and laid on from thirty to forty cart-loads per acre, will laft twenty years. If in plough tilth, keep it up with good rotten dung. If the land be not kind for clover, fummer-tilth for wheat. Small beans, vetches, and grey peafe, will make provifion for wheat, if clean and well-conditioned. Red Lammas wheat is beft for cold lands. Vetches cut green are excellent fodder for horfes—if feeded, they yield from two to two quarters and a half per acre; grey peafe, three quarters: Wheat does well after them.

The above land, if laid down for three or four years until it becomes a thick flag, and then
covered

covered on the flag with forty tons of clay, or twenty tons of marle, or twelve tons of foapers' afhes, per acre, will produce good corn and clover for twenty years.

For gravelly, cold, or wet land, under-drain, if it lie with a proper fall;—by thus removing the caufe, the effect will ceafe. Summer-tilth, and make it clean; lay on from thirty to forty loads of fand per acre, if a little loamy, the better; or about fixteen loads of the above-mentioned compoft, or ten or twelve tons of foaper's afhes, laid on in hard froft, will anfwer well.

4*thly*. We have not difcovered any new manure more efficacious than thofe abovementioned. The burning of clay in kilns has been talked of, but not yet practifed.

5*thly*. In cold wet paftures that cannot be underdrained, make open drains, floped off eafy on each fide; keep them open, and make them with proper falls: then lay on foot, lime, or lime-rubbifh, foapers' afhes, ftreet-dirt, &c. about fourteen loads per acre, and it will laft 14 or 16 years.

6*thly*. Materials for under-draining are, alders and fallows, or ling and black-thorn bufhes, cut and laid in green, covered with peafe or wheat-straw, and above it ftrong clay. Drains thus made will laft twenty years.

7*thly*. The kinds of wood we find to flourifh beft on boggy foils are, alder, fallow, willow, and poplar. Scotch fir does well in a barren foil, efpecially if it has a gravelly bottom.

8*thly*. Thefe graffes are not raifed with us.

10*thly*. We moftly prefer the broadcaft to the drill hufbandry.

11*thly*. We know of no other advantage in the ufe of oxen, than that of keeping lefs ftock; as horfes are more expeditious, and will pay for their keeping by extra labour.

12*thly*. The moft effectual preventative for the rot in fheep is to keep them on dry land; it being found, by general experience, that wet lands bring the rot upon them, efpecially if the feed be bare. In order to cure them, many experiments have been tried, but to little purpofe.

13*thly*. Few new improvements in implements of hufbandry, that are of much confequence, have been made or adopted in this part of the country.

General

General Rules for the Improvement of Lands, by Claying and Marling, as practised by us.

1. Lands that have been many years in plough tilth, and are become foul, may be made clean by a summer-tilth. When this is done, lay on from sixty to eighty tons of clay, or from twenty to thirty tons of marle, per acre. Work it well into the lands, and then sow turnips as before-directed. Feed the turnips off, or at least half: by the treading of the cattle and their manure, the clay will incorporate and work more kindly with the soil. The spring following sow it with barley.——Or,

To clay upon a clover-stubble before the wheat is sown, is a very good method;—it will be fit for a summer-tilth the next year.

2. To improve waste or heath-lands, clay or marle on the flag, from thirty to forty loads of marle, or one hundred and twenty tons of clay, per acre. Turn it in with a good whelming-plough, a moderate depth, in the beginning of February; the sooner the better. If the soil be red and sandy, sow it with white oats. If a black gravelly soil, sow black oats in the middle of March, at least four bushels per acre. As soon as the

the crop is off, fow fix pecks of rye per acre on one ploughing: this will make excellent fheep-feed, and expofe the clay to the winter frofts: then fummer-tilth for turnips; feed them off in March with fheep or other beafts. Such manuring is beft for fuch lands.

If it be a kind of loamy foil, fow barley; if a black gravel, oats. Experience teaches knowledge. Try a lay of clover with the following mixture, viz. clover-feed, and black and white nonfuch. If the lay takes, *fet* red wheat upon it —fuch lands fometimes produce three quarters per acre. When they begin to wear out, improve them by the following method:—lay them down with fuch grafs-feeds as fhall be thought proper, and let them reft for three or four years till they become a flag.

Another method of improving fuch lands is, to lay them up againft winter in round ridges, four furrows on a ridge. Early in the fpring, or fooner if the feafon will admit, turn it back, and make an early fummer tilth; then fow it with buck-wheat, fix pecks per acre, and let clover follow, as above-directed. Forbear feeding it in the fpring, as fuch land will not bear treading.

It will answer to lay it down with any kind of grafs-feeds in the above method, and for a longer time. Let the land be well drained, for wherever the water stands, the clover will decay. Sow buckwheat the latter end of May; it will produce three quarters per acre.

On strong rich lands, clover lays with nonsuch, or any strong flag, *set* wheat, as it will answer far better than sowing it broadcast. Three pecks and a half per acre, set in two rows on each flag, is sufficient. Strike it over with a light harrow bushed. The saving of seed more than pays the planting when wheat is only 5s. the bushel; the price of planting from eight to ten shillings per acre.* Small tick beans, 7s. per acre planting; hoeing 6s.; produce three quarters and a half to four quarters per acre. Windsor tick beans, 7s. planting; 6s. hoeing; produce four to five quarters per acre.

* The expence is now reduced to about six shillings per acre.

ARTICLE

Article XIV.

On the Use and Effect of Marle *in Norfolk.*

[By a Gentleman Farmer in that county.]

Gentlemen, March, 1778.

In answer to your enquiries respecting the use of marle in this county, our farmers seldom lay it on pasture, but constantly on arable land, from thirty to eighty, and, in some instances, to one hundred loads per acre. By a load, I mean as much as a cart and three horses can draw. They prefer laying it on a clover and ray-grass, or a barley stubble, or layer, a year before it is ploughed in. By this means it is more intimately mixed with the upper part of the soil, and will not be so soon buried by the plough, as when laid on and turned in immediately.

The marle mostly found with us is, a white pure calcareous substance like chalk, but fat and unctuous. When it is met with of any other colour, our farmers will scarcely be persuaded it can be marle. This I experienced a few years since, upon discovering in my park a fine light brown, or rather dove-coloured marle, with every other property like the white.

The effects of marle have been very great indeed in this county, having advanced the rent of lands upon which it has been laid, in some instances, from half a crown to ten shillings an acre and upwards. This improvement has been chiefly made on light sandy soils. But marle has been found beneficial on all soils. The general opinion with us is, that it not only gives tenacity to the soil, but acts also as a manure by virtue of its salts. Our farmers, after the first dressing of marle alone, mix it with dung or compost, and think it much improved thereby.

When I say *our* farmers, I mean the Norfolk farmers, for I do not know that there is one marle pit within two or three miles of me; or that any marle has been discovered within that distance, except what I mentioned to have found in my own park, which was applied wholly on the grass thereof, being no more than what was dug out of a fosse I was making.

Our use of lime is trivial; and no great benefit has been found to result from it here: but this is entirely owing to its having been used in such small proportions as could not possibly have much effect. This, however, is no argument against the use of

lime:

lime: by a like management, dung, or any other manure, would prove equally ufelefs.

Your idea of the turnip hufbandry is perfectly juft; but an effectual method to prevent the ravages of the fly remains, and I fear will ftill remain, a defideratum in hufbandry.

One obfervation, made by our farmers, I have generally found to be juft; to wit, that the mifchief is greateft in the midft of the land; and have frequently found, that when every other part of the crop was deftroyed, two or three ridges next the hedges have efcaped all injury. This, with fome other particulars, carefully attended to, may, hereafter, lead to fome method of preventing the evil.

I am, &c.

ARTICLE

ARTCLE XV.

On Feeding Wheat in the Spring with Sheep.

[By a Gentleman Farmer in Essex.]

GENTLEMEN,

I Now comply with your request in giving you my thoughts on the practice of feeding off wheat with sheep in the spring, and also an account of my success therein last spring.

This practice ought not to be generally adopted, even where the crop is rank, or, as we term it, *winter-proud*; because, in many cases, it would not answer any good purpose, but, on the contrary, injure the crop. In some instances, however, it has succeeded, and the advantages are these:—it affords feed for ewes and wethers when turnips are over, and before the spring feed comes in; it causes the wheat to branch out into a greater number of stalks than it would otherwise do, and, of course, the crop is increased: by the warmth of the sheep when lying on it, and the manure they leave, the crop is brought forwarder, and the grain heavier than it otherwise would be. In light lands especially, the treading of the sheep fixes the earth about the roots of the corn, and causes the ground to retain its moisture longer in a dry spring.

On the other hand, this practice is liable to the following disadvantages:—In some lands it checks the growth of the corn, and makes the second shoots weak and small—of course the ears will be small, and the grain light in proportion;—in foul lands and a wet season, it gives opportunity for the weeds to rise above the corn so as to choak it. Sheep are also apt to bite off the knot of the plant.*

This practice answers best on clean land, and a light soil. Here the treading of the sheep is of service; and there is no danger of the weeds rising so as to injure the crop.

In September 1777, I sowed fourteen acres of wheat, which, soon after Christmas, seemed winter-proud. The soil was a loose loam, and I had laid on plenty of dung. In the beginning of February, I turned about sixty sheep into the field, and fed it down; but the weather coming in milder than I expected, the weeds produced by the dung got so much a-head of the wheat, that the crop was a very poor one—not more than nineteen bushels per acre.

* To prevent this, the Farmer should turn them in hungry, and take them out as soon as they have filled their bellies. When hungry, they will eat the leaves of the plant; but when their hunger is satiated, they will pick out the knot or crown of the plant, that being the most sweet and delicate.

I had

I had another field of wheat, which was poor land, and being a turnip fallow, was clean, and had not been manured. In February, I observed the plants to be small, and to stand thin, and therefore turned in some sheep, thinking it would cause the plants to throw out more side stems. The knot of the plants not being much above ground, there was no danger in that respect: the experiment succeeded, and I reaped near four quarters per acre.

From the closest observation, I find that wheat ought not to be fed down with sheep, unless it be very rank in January; and that such only should be fed as was sown early on land that is neither rich with dung nor weedy.

After it is fed, if the land be clean, a top-dressing of soot, ashes, malt-dust, &c. will greatly cherish the crop. I have experienced this in many instances, and can safely recommend the practice.

I am, &c.
P. W.

Jan. 24, 1779.

ARTICLE

Article XVI.

Method of making Ponds in dry Countries, for Watering Sheep and Cattle.

[Communicated by a Gentleman near Beverly.]

MARK out a circular piece of ground, whose diameter is twenty yards, (more or less) and dig out one foot of earth, so as to leave the sides perpendicular that depth. Then begin to form it in the shape of a wooden milk-bowl, till the perpendicular depth in the centre be four feet and a half or five feet. On the bottom and sides spread lime, finely powdered, two or three inches thick. On this lime lay well-tempered clay, six or seven inches thick. This clay, when laid on, must be well worked with circular beaters of a foot diameter and three inches thick, first using the outside edge of the beater, which will indent the clay, then use the flat side, so as to leave it with a smooth surface. Upon the clay thus prepared, lay gravel or chalkstone six inches thick. The gravel should have both the finer and coarser parts screened from it. No more clay should be prepared for the gravel than can be laid and covered the same day, as heat or frost will be equally apt to catch it, which must be particularly guarded against, as it would occasion

the

the pond to lose its water. After the gravel is laid in, nothing more is necessary.

A piece of ground should be chosen for this purpose, to which there is a descent from all sides, if it can be found in a proper situation.

Winter, or early in the spring, is the best season for making these ponds or reservoirs.

Lay each material of equal thickness from the centre to the edges of the pond.

If lime can be made fine enough without the use of water, so much the better: if not, use as little water as possible. The clay should have no more water than will serve to make it work kindly.

In this manner ponds may be made of any size, the diameter and depth being kept nearly in the same proportion as above-directed.

After I left Bath the last summer, and before the end of the long drought, I saw in a field one of these ponds nearly two-thirds full of water, although many cattle, sheep, and horses, had grazed there since the beginning of May.

I am, &c.

Jan. 27, 1779.

Article XVII.

Experiments on Plants eaten or rejected by Cattle, Sheep, and Hogs, recommended.

Gentlemen;

THOUGH the use of botanic science has been principally restricted to medicine, yet it certainly has a natural and inseparable connection with agriculture; some of the most important branches of which depend on the knowledge of it, particularly that which respects the feeding of cattle.

That Agriculture has not been studied, or encouraged in proportion to its great importance, or advanced with equal rapidity as divers other arts or branches of knowledge, is a fact generally admitted.

From an indubitable conviction of this truth, we may date the origin of those publick institutions for its support and improvement which reflect honour on our age and nation.

The commendable spirit which appears in the Society instituted at Bath, furnishes encouragement to hope that agriculture will receive much additional

tional improvement in the Western counties; and as a well-wisher to the undertaking, I beg leave to throw a few remarks before you on a subject not unworthy your attention.

It is well known that grasses furnish the principal food of our cattle; but among the natural classes of plants, there are many, of the leguminous tribe especially, on which they feed with avidity.

Numerous instances, however, occur of one class of animals feeding eagerly on those plants which others refuse to touch. Plants, that are noxious and even poisonous to some animals, are freely eaten by others without the least inconvenience.— Hence it seems that there is a peculiar structure in the vessels of each species of beasts, to which only the particles of different vegetables are respectively adapted. But there has not, to my knowledge, been any regular course of experiments made in England for ascertaining precisely the several species of plants thus eaten or rejected, or a regular list formed and published of those that are noxious. Such a course of experiments is greatly wanted, and would doubtless be productive of much benefit to the publick.

The celebrated Linnæus superintended a great attempt of this kind in Sweden many years since, the result of which may be seen in the second volume of Amœnitates Academicæ, and is, I think, highly deserving your attention.

It was in his Dalekarlian Journey that Linnæus conceived the first design of this great work. In that tour he found that his horses left untouched, among other plants, the following: Meadow-sweet, great wild valerian, lily of the valley, angelica, rose-bay willow, marsh cinquefoil, mountain and globe crowfoot, cranesbill, yellow wolfsbane, and several shrubs. Soon after his return, he and some of his pupils set about the work. Above two thousand experiments were made on horned cattle, sheep, horses, hogs, and goats, with the sole view of determining what kinds of vegetables those several animals would eat or reject.

These experiments being made with great care and accuracy, the result of them must on the whole be true and conclusive; as it has a real foundation in that unerring law of instinct, established by the God of nature in the whole brute creation.

As these experiments were made on the indigenous plants of Sweden, they can only be decisive here with respect to plants common to both countries; but as they take in the greater part of English

lish plants, they would greatly facilitate an attempt of the same kind in this kingdom.

Three-fourths of the plants growing with us are the same as those in Sweden on which experiments have been made. One fourth only remains to make new experiments upon. The undertaking would, therefore, not be so great or difficult to accomplish, as at the first view it may appear to be. The country round Bath, consisting of great variety of soil and surface, is particularly favourable for such a work. Fenny ground and the sea coast are not too remote to be visited on the occasion.

The advantages arising from this course of experiments being rendered complete, would be important and lasting. Poisonous and noxious plants might be eradicated. The farmer would know with certainty what to cultivate and what to reject. Fen lands might be rendered valuable by the introduction of plants suited to the soil. By improving the produce of pasture land, our hay would be finer and better; and in proportion as the food of cattle is purely what instinct points out to them, their flesh must be finer, and better adapted to human food, than when supplied with a mixture of juices of an unfriendly or noxious quality.

I am, &c.   X. Y. Z.

ARTICLE

Article XVIII.

On the Bulk and Increase in Growth of some remarkable Timber-Trees.

GENTLEMEN,

I Herewith send you inclosed a letter which I lately received from a most worthy and ingenious gentleman of this county, whose accuracy and fidelity cannot be doubted; the subject of it is both curious and entertaining, whilst it will at the same time afford an opportunity of drawing from it conclusions the most useful and interesting to the public. You will perceive, that I am permitted by my friend to communicate it to your truly patriotic Society; whose condescension in having honoured me with the title of a Member of it, as it claims every mark of respect, and every attention I can possibly pay to their views and inclinations, so it would have left me without excuse, had I omitted adding this valuable performance to their collection, which will, I presume, be enriched by it.

I am, with particular respect, your's, &c.

THOMAS BEEVOR.

Hethel, near Norwich,
 Oct. 11, 1779.

'DEAR

'DEAR SIR,

Stratton, Oct. 1, 1779.

'IN compliance with your request, I here send you the measures of some of the largest trees, taken by myself, in several rambles about the kingdom. But although I have been in parts of every county of England and Wales, perhaps larger than these may have escaped my search; as I never heard of the Demary oak by Blandford, until I read Mr. Hutchinson's account of it in his history of Dorsetshire. The largest oak I have seen is that at Cowthorp, or Coltsthorp, near Wetherby in Yorkshire, of which the ingenious Dr. Hunter gives a plate in his Edition of Evelyn's Sylva. The Doctor calls this tree 48 feet circumference at 3 feet from the ground; and I found it in 1768, at 4 feet, 40 f. 6 in.; and at 5 feet, 36 f. 6 in.; and at 6 feet, 32 f. 1 in.—Here to save repetition, 5 feet is the height I always measure at, as easier to see the level of the string, and also being clearer of the swellings of the roots.

'In 1759, the Oak in Holt Forest, near Bentley, was, at 7 feet, 34 ft. There is a large excrescence at 5 and 6 feet, that would render the measure unfair. In 1778, this tree was increased half an inch, in 19 years. It does not appear to be hollow, but by the trifling increase, I conclude it not sound.

The

'The Fairtop Oak in Epping Foreſt, ſeeming ſound in 1754, and the Earl of Thanet's hollow Oak, in Whinfield-park in Weſtmoreland, in 1765, were both 31 f. 9 in.

'The handſomeſt Oak I ever ſaw was in the Earl of Powys's noble park by Ludlow, in 1757, though it was but 16 f. 3 in. But it ran quite ſtrait, and clear of arms, I believe, full 60 feet, and had a large and fine head.

'In Benel church-yard, 3 miles north of Dunbarton, in Scotland, in 1768, a very flouriſhing Aſh, 16 f. 9 in.

'In 1754, a fine Wych Elm by Bradley church, in Suffolk, 25 f. $5\frac{1}{2}$ in. In 1767, this tree was 26 feet 3 in. Increaſed $9\frac{1}{2}$ in. in 13 years.

'I have a hollow Wych Elm by Stratton church, at 4 feet, 29 f. 6 in. and I had in 1760, in my old park in Hevingham, a headed Alder, at 4 feet, 16 f. $2\frac{1}{4}$ inches.

'In 1755, your Hawthorn, by Hethel church, was, at 4 feet, 9 f. $1\frac{1}{2}$ in.; and one arm extended above 7 yards.

'The

'The talleſt trees that I have ſeen were Spruce and Silver Firs, in the vallies in Switzerland. I ſaw ſeveral firs in the dock-yard in Venice above 40 yards long; and one of 39 yards was 18 inches diameter at the ſmall end. I was told they came from Switzerland.

'In Lord Petre's old park, at Writtle in Eſſex, in 1764, I found a Hornbeam above 12 feet; and the old Cheſnut, (very hollow) at $3\frac{1}{2}$ feet, the leaſt part 42 f. 5 in.; at 5 feet, 46 f. 1 in.; and at 6 feet, 49 f. $5\frac{2}{4}$ inches.

'In 1759, the Cheſnut in Lord Ducie's garden, at Tortworth in Gloceſterſhire, was at 6 feet, (the loweſt I could meaſure it, as the garden-wall joins to the tree on two ſides) 46 f. 6 in.; it did not appear hollow, but had very few and ſmall boughs: as I took the meaſure in a heavy rain, and did not meaſure the ſtring till after I returned to the inn, I cannot ſo well anſwer for this, as the other meaſures.

'I omit beech, birch, maple, abele, &c. as I have heard of much larger trees of thoſe ſorts than I have ſeen.

'Perhaps an account of the annual increaſe of ſome trees will not be foreign to your purpoſe.—
You

You know it is difficult to discover the age of old trees, as very few old planters kept registers of their plantations.

' I have seen a memorandum of a former Rector of Hevingham, wherein is written, that "in 1610 he planted two chesnuts by his church porch;" the largest was, last autumn 1778, 14 f. 8¼ inches, or 176¼ inches in 168 years. Supposing the tree to have been 9¼ inches when planted, you see it increased an inch yearly.—And I have a deed between an ancestor of mine, as lord of the manor of Stratton, and his copyhold tenants, upon his inclosing some of the waste, wherein the abuttal to the west is upon the road leading from Hevingham to Norwich, which you know cannot be mistaken: the date is 1580, and the largest oak on that bank, at 4 feet, was, last autumn 1778, 16 f. 3½ inches, or 195½ inches in 198 years.

' Now, from the increase of the Bentley oak, and the two last-named trees, I conclude the Tortworth chesnut is not less than 1100 years old; perhaps it may be much older.

' I offer you the following calculation for your amusement; from its vast bulk, you must conclude it was a very healthful tree. Suppose it increased

an

an inch and quarter yearly the first century, an inch the second, three quarters the third, half an inch the fourth, one-third the fifth, and thirty inches each century for the second 500 years, and a little less than a quarter for the eleventh century; the account will stand thus:

	Inches.
The first century, at an inch and quarter —	125
Second ditto at one inch — —	100
Third ditto at three quarters — —	75
Fourth ditto at half an inch — —	50
Fifth ditto at one-third of an inch —	33⅓
Second 500 years at 30 inches per century —	150
Eleventh century at 24¾ inches — —	24¾

46 feet and a half; or 558 i.

'There is a tradition, that this tree was called the Great Chesnut in King John's time; and supposing it grew in this proportion, it was 540 years old when he came to the throne, and eleven yards in circumference.

'Sir R. Atkins, in his history of Glocestershire, (p. 413) says, 'by tradition this tree was growing in King John's reign, and is 19 yards in compass;' and I believe it is at least so large near the earth. Now, although I have sufficient proof of young trees increasing much more than my supposed growth

growth of this chesnut, yet perhaps I have allowed sufficiently for that tree, as it grows on a stiff clay; which, though perhaps it may in the end produce the largest trees, yet I believe most trees will grow faster in lighter soils.

I planted an oak in 1720, which was last autumn 7 f. 9 in. I do not pretend to remember the size when planted; but in autumn 1742, it was 2 f. $11\frac{1}{4}$ in.; *i.e.* $57\frac{3}{4}$ in. increase in 36 years,—above an inch and a half yearly. But this oak was taken from very poor land to a tolerable light soil, and stands single; and perhaps the growth was helped by digging a large circle round it in several winters, and in other years having that circle covered with greasy pond mud; and in some dry seasons, I washed the stem: the advantage of washing I experienced in 1775, greatly to my satisfaction. You may see the full account in my letter to the Bishop of Bath and Wells, in the 67th volume of the Philos. Transf. in 1777. But supposing these endeavours did not help the growth of this oak, yet I apprehend it will not be 225 inches in circumference when 200 years old. For though the Hevingham chesnut is a healthful tree, it has increased but 25 inches and a half in the last 36 years, (viz. from my first measuring it) which shews, if it had not gained more in its younger state, it would have

taken

taken 250 years to make its present bulk of 176 inches: and my oak of 198 years old has, from 1760, increased only 12¾ inches in 18 years: which proportion would take 275 years to make 195 inches: and the oak by Bentley, according to the last 19 years increase, would take above 15500 years to make 408 inches, the present circumference of the tree.

'If you think any of these measures will afford entertainment to the Society at Bath, you have my leave to offer this letter to them; which will shew at least that I have pleasure in obeying your commands.

'I am, with great respect,

'Your most humble and obedient servant,

R. MARSHAM.'

'*Thomas Beevor, Esq.*'

P. S. I have put dates to all the measures, that if your curiosity should lead you to measure any of the trees, you may know what progress they make in a certain time.'

ARTICLE

Article XIX.

Mode of Farming pursued by a Member of the Bath Agriculture Society.

GENTLEMEN,

WHEN I took my farm, which consisted of 115 acres, I found the arable undone by improper tillage, and the meadows worn out for want of manure. I will describe the farm as nearly as I can. The soil in general is composed differently, of mould, sand, gravel, and here and there clay. The farmer I succeeded was a sloven in the abstract, and so bad a ploughman, that he never cross-ploughed his land through incapability. It was an opinion of mine, that the sooner I got my land in order the better; and that the first expence would be lightest.

No. I. was a field lying near the house, tolerably clean and not wanting dung, found in a barley-stubble without clover; this I planted with pease, and giving it a dressing of dung, I had a good crop. This was the year 1775, and the succeeding year I had a good crop of wheat, not indeed equal to my improved land, but little less than three quarters per acre. The year following the wheat, I summer-
fallowed

fallowed and turniped it,—the courſe I afterwards invariably purſued. The ſoil was gravelly, with ſome depth of mould.

'No. II. was a coarſe unkind piece of land, of a ſoil neither clayey nor gravelly, but ſomething between both, and which my men called chiſley. This had been cropped with oats, and a very indifferent crop indeed; I dunged it in the ſpring, and planted potatoes on half, and ſowed vetches on the other half; the crops of both were equal to my expectation, but the greateſt advantage was the benefit the land received from the potatoes, by which it was mellowed ſo ſurpriſingly, that I was reſolved, contrary to my firſt intention, to ſow the field with wheat, which yielded me only four ſacks per acre.

No. III. had borne wheat, but the ſtubble was ploughed up for turnips, which afforded a little ſheep-feed, though not worth the expence. This was a good piece of land, tolerably clean, but wanted reſt. The ſpring turning out favourable, I altered my intention, which was to have fallowed it and ſown it with turnips, and therefore ſowed it with barley at five ploughings; with the barley,' three buſhels per acre, I ſowed a buſhel of rye-graſs, 6 lb. of Dutch clover, and 6 lb. of black-graſs, without

out any broad clover; not that I know it to be a good method, but that I wished to lay it down for some years, being handy for feeding. The barley yielded me about $3\frac{1}{2}$ quarters per acre. As soon as the barley was off, I dressed this field with chalk and compost of dung and earth separately, the chalk about eight waggon loads per acre, and the compost about twelve cart loads; I had the year following a most excellent swarth of grass.

No. IV. had been cropped with beans. This I sowed with wheat at three ploughings. After the wheat I had it tilled and dunged for summer vetches, of which I had a most noble crop. When the vetches were off, I had it ploughed three times, and sowed with wheat.

No. V. a clover-lay, I sowed with wheat at one ploughing; it was very foul, and produced only about three sacks per acre. The year following it was turniped.

No. VI. was a rowetty coarse piece of pasture, that had not been ploughed for some years. Here I sowed black oats; the produce was three quarters and a half per acre; the next year it was summer-fallowed and sown with turnips.

No. VII.

No. VII. a mixture of fandy and gravelly foil, was from a wheat-ftubble fummer-fallowed, well dunged, and fown with turnips, as was alfo

No. VIII. which was drained and chalked. The turnips on both were remarkably good. No. VII. was the next year turned into a garden, and bore very large crops of potatoes, Windfor beans, carrots, cabbages, and parfnips, but a very few onions; which I apprehend was as much owing to want of care as any thing elfe; among the beans I tried fome turnips, but they were not good, being in general worm-eaten, and fticky or ftringy.

No. IX. part of which was an orchard, was alfo in grafs: this I dunged well, and planted with beans, thinking to fummer-fallow for turnips the next year, but in this I altered my mind, and fowed winter vetches, which anfwered moft incomparably well, cut for horfes, and for feeding the fows and pigs. The beans produced rather more than four quarters per acre: they were hoed three times, which did them great good, but deftroyed no couch. After the vetches, I fummer-fallowed for turnips.

No. X. was a meadow which had been ftrangely neglected; a brook ran through it, and frequently over-

overflowing, had given nurture to abundance of rushes. I ordered water-furrows to be cut, sufficient to carry off all wet, and spread over the whole meadow wood-ashes brought dry from a lime and brick-kiln, to the quantity of forty bushels per acre; I kept it close fed that summer instead of mowing it: in the winter, a little before Christmas, I dressed it well with dung, and the produce of grass answered well the expence: I cut full two tons per acre of excellent grass free from rushes.

My aim was to bring all my land as soon as possible to bear turnips, which I consider as the foundation of good husbandry, in the following order: Turnips, barley, clover, and wheat, and this succession invariably. After I got my land in order, for which I spared no expence, my crop was large, five quarters of barley, and from eight to ten sacks of wheat in general. I made it a rule always to manure my clover as soon after the barley was off as I could; and this dressing was of the best materials I could collect, with stable dung if I could get it. But to proceed in my story with the rest of my farm; for I have only yet mentioned fifty-two acres.

No. XI. had borne a second crop of oats self-sown: this, to be sure, must be summer-fallowed. I had

I had it ploughed ſix times, and manured with 12 good waggon-loads of ſtable-dung well rotted to an acre; but being ſtoney land, I ſowed it with wheat inſtead of turnips, and yet the produce was not more than three quarters per acre. I think, from the experience I have had, that the dung is not of very eſſential ſervice when applied for a crop of corn, but to turnips, pulſe, graſs, or vetches, it is of the utmoſt importance; and after theſe crops will wonderfully aſſiſt the corn crops, as barley after turnips, and wheat after peaſe, vetches, or clover.

No. XII. was a clover lay, which; not having been dreſſed for graſs, I manured for wheat, but the produce was very ſmall, not more than three ſacks and a half to an acre; this was turniped after the wheat.

No. XIII. XIV. and XV. I ſummer-fallowed, well dunged, and ſowed with turnips; theſe were a light lively land, capable of being worked after a month's rain, and yet not burning. The turnips were remarkably good, the barley five quarters per acre. As ſoon as the barley was off, the clover was dunged; and the produce of the clover, at two cuttings, three tons and a half per acre. The wheat was ſown at one ploughing, two buſhels per acre,

acre, and the produce full nine sacks and two bushels upon the average.

No. XVI. was a barley stubble, with a good plant of clover. I dressed the clover well, and mowed near two tons at two crops, and sowed the land with wheat. In the spring the wheat was very thin and worm-eaten. I sowed some soot over the parts that were injured, which stopped the further progress of the worm; and the land being in good heart, from the dung I had put on the clover, the wheat tillered amazingly, and produced, totally unexpected by me, three quarters and a half per acre. However a spring cleaning of foul land for barley may answer for that crop and the crop of grass, yet when the land comes to be sown with wheat afterwards, the couch will almost get the better of the wheat, and inevitably do it considerable damage; of this I had an instance in No. XVII. which was a clover-stubble left after barley, where great pains had been taken to clean the land and rid it of couch; but the land, when turned up and sown with wheat, was so foul, that the crop hardly paid the expences, and I repented I did not summer-fallow and sow it with turnips out of the clover.

No. XVIII. I sowed after pease with white oats and clover, but the clover did not take kindly; and

as

as the oats were got off pretty soon, my man advised me to sow it with wheat; accordingly I had it ploughed four times, and got out all the clutter of couch, weeds, &c. that we could; after that I dunged and sowed it with wheat: the crop was not very much amiss, though not equal to what I expected, and I might better have turniped it at once.

No. XIX. was wheat stubble, which I summer-fallowed and sowed with turnips.

Thus I have given the method of farming I pursued in Berkshire without imposition or exaggeration.

No.	Soil.	Acres.	1775.	1776.	1777.	1778.
1.	Gravelly mould	6	Pease	Wheat	Turnips	Barley
2.	Gravel and clay	6	Potatoes, Vetches	Wheat	Turnips	Barley
3.	Deep loamy mould	4	Barley	Grass	Ditto	Ditto
4.	Stiff soil	5	Wheat	Vetches	Wheat	Turnips
5.	Mellow loam	4	Wheat	Turnips	Barley	Clover
6.	Sandy loam	4	Oats	Turnips	Barley	Clover
7.	Sandy and gravelly soil	4	Turnips	Potatoes, &c.	Wheat	Turnips
8.	Ditto	6	Turnips	Barley	Grass	Ditto
9.	Mellow loam	3	Beans	Vetches	Turnips	Barley
10.	Meadow	10				
11.	Strong clayey soil	5	Sum. fal.	Wheat	Beans	Turnips
12.	Stiff clayey gravel	5	Wheat	Turnips	Barley	Grass
13, 14, 15.	Sandy loam	12	Turnips	Barley	Clover	Wheat
16.	Ditto	4	Wheat	Turnips	Barley	Clover
17.	Sandy & gravelly loam	5	Wheat	Turnips	Barley	Clover
18.	Gravelly soil	6	Wh. Oats	Wheat	Turnips	Barley
19.	Sandy & gravelly loam	10	Turnips	Barley	Clover	Wheat
20, 21.	Meadow	16				

By this mode of farming, I had only nine acres of wheat in the year 1777; a considerable less quantity than any other year. But in general, I wished to have the farm as equally divided as possible into the several crops of turnips, beans, clover, and wheat.

T. L.

Article XX.

On the best Method of raising Elms for Fences; manuring Fallows for Wheat; and preventing the ravages of the Fly on young Turnips.

Gentlemen,

THE best method of raising elms quick is the following: When you fell elm timber, in the spring, sow the chips made in trimming or hewing them green, on a piece of ground newly ploughed, as you would corn, and harrow them in. Every chip which has an eye, or bud-knot, or some bark on it, will immediately shoot like the cuttings of potatoes; and the plants thus raised having no tap-roots, but shooting their fibres horizontally in the richest part of the soil, will be more vigorous, and may be more safely and easily transplanted, than when raised from seed, or in any other method. For

For elm fences, the plants thus raised have greatly the advantage of others, as five or six, and frequently a greater number of stems will arise from the same chip; and such plants, when cut down within three inches of the ground, will multiply their side shoots in proportion, and make a hedge thicker, without running to naked wood, than by any other method yet practised. If kept clipt for three or four years, they will be amoft impenetrable.

Stable-yard dung is commonly used as a manure, on land intended to be sown with wheat; but let it be observed that this dung is more productive of weeds than any other manure. A crop of wheat cannot be kept too clean; hence much trouble and expence are occasioned by so injudicious a process. To remedy or rather to prevent this inconvenience, instead of sowing a newly dunged fallow with wheat, sow it first with white oats; these will take off the rankness of the dung, destroy numberless weeds, and leave the land in excellent order for wheat the following autumn. In short, it is nearly equal to a turnip-fallow.

The ravages of the fly on turnips have frequently occasioned great loss to the farmer, and many remedies have been proposed, most of which have

not

not anfwered—perhaps the following may be more efficacious: The boughs of the common elder-tree, fixed in a gate, and drawn gently over young turnips when they firft appear, will prove an excellent prefervative from the fly; and if the leaves of the faid boughs be a little bruifed, and fumigated with the fmoak of burnt tobacco mixed with a fmall quantity of affafœtida, it will deftroy thofe infects effectually. It will alfo be of great ufe to brufh the leaves and branches of your wall-fruit trees with elder boughs thus prepared. Nothing is fo difagreeable to infects as a mixture of tobacco and affafœtida fumigations. It will kill them inftantly wherever applied.

I live at too great a diftance to attend your meeting, but wifh to promote your laudable defigns as far as I am able.

I am, &c.

Taunton-Dean, July 4, 1778.

To the above letter, we think it not improper to fubjoin an extract from another, fince received from a gentleman at Exeter; and recommend the experiments mentioned in both as a remedy for the fly in turnips.

'After

'After the land is ploughed for turnips, and
'when the seed is harrowing in, let some large
'branches of common elder, with the berries on,
'be fixed in the harrow, so as to rub on the ground.
'The friction of the leaves and berries will leave so
'strong (and to these insects, so disagreeable) a taint
'or odour on the soil, as will probably prevent
'their alighting on so unpleasing a spot, or make
'them speedily leave it, if they can be supposed to
'have been there before the seed was sown. The
'effects of the effluvia of elder are much greater,
'and more lasting, with respect to those insects,
'than would at first be imagined, or even credited
'by the bulk of mankind.'

ARTICLE XXI.

On a peculiar species of Grass found at Orcheston, on Salisbury-Plains, Wiltshire.

[By a Gentleman of Dorchester.]

GENTLEMEN,

I AM favoured with your Secretary's obliging letter, in reply to mine respecting the grass-seed; and it gives me satisfaction that I can herewith send you a specimen in the blade for your inspection. This grass is found at Orcheston St. Mary, about nine

nine miles from Salisbury, in a meadow belonging to Lord Rivers, now in the occupation of Farmer Hayward. This meadow, being situated on a small brook, is frequently overflowed, and sometimes continues so a great part of the winter. It bears the greatest burthen in a wet season.

When I was there, it was too early in the spring to make any particular observation on the blade, but the Farmer's account is as follows, viz. ‘ that it generally grows to the height of about eighteen ‘ inches, and then falls and runs along the ground ‘ in knots, to the length of fifteen or eighteen feet, ‘ but that he has known instances of its running to ‘ the length of twenty-five feet.’

The meadow contains about two acres and a half. It is mowed twice in a season, and the average quantity is generally about twelve loads (tons) of hay the first mowing, and six the second; though sometimes considerably more. The tithe of the meadow has been compounded for at nine pounds a year.*

The

---

* This account appeared to us so singular, and the crop of grass so very extraordinary; that our Secretary went to Orcheston, to examine more particularly into it. The farmer, and divers other persons in the village, confirmed the account contained in this letter, of its amazing produce in summers when the meadow had been overflowed in the preceding winter and spring; but when the winter had been dry, and the

The grass is of a sweet nature; all cattle, and even pigs, eat it very eagerly. When made into hay, it is excellent, and improves beasts greatly. The farmer says his horses will eat it in preference to corn mixed with chaff, when both are set before them together.

Should the Society wish for further information or assistance, I shall be happy in doing every thing in my power to promote their views.

the meadow not overflowed, the crop of grass was not near so large. There did not appear to be any thing peculiar in the soil; nor were the other plants or weeds growing on it more luxuriant than in many other similar situations. Some of this grass was sent to the Society at Norwich; some ingenious members of which inform us, that they think it is a species of the Agrostis Polymorpha, mentioned by Hudson in his *Flora Anglica*, of which there are several varieties.

Camden mentions, in his *Britannia*, a grass growing near the place where this is found, which he calls *trailing Dog's-grass*, and says that "hogs were fed with it."

From all the enquiry made, we have not found this species of grass growing in any other part of the kingdom; hence it is possible that there may be something in the soil of this meadow peculiarly favourable to its growth.

We shall not, however, determine on this point, but recommend trials to be made of propagating it, by sowing the seed in other places, subject to be overflowed in the same manner. If it can be propagated generally, it must turn out the most profitable to the farmer of any grass yet discovered, and be of great benefit to the community.

ARTICLE

Article XXII.

Some Observations on Thistles *as injurious in Agriculture, more particularly the* Seratula Arvensis *of* Linnæus.

[By W. Curtis, Author of the *Flora Londinensis*.]

Gentlemen,

WHILE some of your correspondents are laudably engaged in enriching agriculture, by discovering and promoting the cultivation of new plants, permit one whom you have been pleased to elect an honorary member of your Society, to lay before you a few observations on some of the plants which are more particularly noxious to the farmer. Should they be considered as contributing to advance even in the smallest degree the design of your institution, he may be again excited to trouble you on other subjects, as information may arise from a cultivation of most of the British plants on a small scale.

There are no plants over which the œconomical farmer ought to keep a more watchful eye than the thistle tribe. He is sensible that they are not only useless, as resisting the bite of most animals, the hardy ass excepted, but that they occupy much ground; and being furnished with downy seeds, are

capable

capable of being multiplied to almoſt any diſtance. Hence in many parts of the kingdom, the farmers whoſe lands are contiguous unite in preventing the increaſe, by cutting them down before they feed; but this operation, though deſtructive to ſome ſpecies, will only palliate the bad effects of others.

To be acquainted with the qualities of each kind, we muſt obſerve them with much attention, and view them in a botanical and philoſophical light: this alone will enable us to judge with certainty how far and by what means their deſtruction may be effected.

The Engliſh Thiſtles meriting notice, as more or leſs noxious, are,

1. The Carduus Lanceolatus, or Spear Thiſtle
2. Carduus Nutans, — Muſk Thiſtle
3. Carduus Paluſtris, — Marſh Thiſtle
4. Carduus Marianus, — Milk Thiſtle
5. Carduus Acanthoides, — Welted Thiſtle
6. Carduus Criſpus, — Curled Thiſtle
7. Onoperdum Acanthium, — Cotton Thiſtle
8. Seratula Arvenſis, —*Curſed Thiſtle

* The term *Way Thiſtle*, by which this plant has uſually been diſtinguiſhed, is by no means ſufficiently expreſſive of it. The preſent term may, perhaps, be thought too harſh; but if any plant deſerve to have a mark ſet upon it, it is certainly this.

The

The *Spear Thistle* is a large strong plant, about four feet high, the extremity of each leaf running out into a long point; its heads are large, and it grows very commonly by the sides of roads, near dung-hills, and not unfrequently in fields and pastures.

The *Musk Thistle* grows to the height of 2 or 3 feet, the heads hang down, and the flowers smell somewhat like Musk; it is often found occupying whole fields, particularly on chalky or barren land.

The *Marsh Thistle* is very tall and prickly; its heads are numerous, small, and of a red colour; it grows abundantly in wet meadows, also in woods.

The *Milk Thistle* has very large leaves, which are most commonly beautifully marbled with white. Near London it appears frequently on banks by road sides; in which situation we also meet with

The *Curled* and *Welted Thistle*. These three seldom intrude into fields and pastures.

The *Cotton Thistle* is distinguished by its size, (being perhaps the largest of the British herbaceous plants) and its white woolly leaves. It grows in the same situation as the three last-mentioned.

The

The *Curfed Thiftle* is more general in its growth than any of the others, being found not only by the sides of roads universally, but also in arable land, and is not uncommon in meadows, even in such as are yearly mown. It is remarkably prickly, grows about three feet high; its heads are small, the flowers purple, and frequently white. The scales of the heads are smooth, and may in a particular manner be distinguished from all the others before-mentioned, by having a perennial root about the size of a goose-quill, which runs deep into the earth, and afterwards creeps along horizontally.

Of these thistles, all except the last are either *annual* or *biennial*; that is, remain in the ground not more than one or two years, unless renewed by seed. The last, having a perennial root, continues in the earth, increasing and throwing up new shoots every year.

Hence it will appear obvious, that if the first seven species of thistles are cut down before they perfect their seed, the ground will be entirely cleared of them; and that the last-mentioned can no otherwise be destroyed, than by rooting it out, a process which the following experiments will sorrowfully convince the rural œconomist to be impracticable

in

in large fields, and scarce to be performed even in an inclosed garden.

EXPERIMENT FIRST,

To ascertain the effects of mowing the Seratula Arvensis.

The Hon. Daines Barrington, who is ever anxious to promote useful enquiries, desired me to try whether this kind of thistle could not be destroyed by mowing. A small patch of them, about two feet square, was accordingly planted in a good garden, in the year 1777. In the course of the summer they were mown three several times, but without any other good effect than that of preventing their seeding: for instead of being destroyed, the next spring they came up extremely vigorous, not only on the bed where they were first planted, but all around it to the distance of six feet.

EXPERIMENT SECOND,

To ascertain the annual Increase of the Root of the Seratula Arvensis.

April 1, 1778, I planted in a garden a piece of the root of this thistle, about the size of a goosequill, and 2 inches long, with a small head of leaves, cut off from the main root, just as it was springing

out

out of the ground. By the 2d of November, 1778, this small root had thrown out shoots, several of which had extended themselves to the distance of eight feet; some had even thrown up leaves six feet from the original root. Most of these shoots which had thus far extended themselves were about six inches under ground—others had penetrated to the depth of two feet and a half; the whole together, when dug up and washed from the earth, weighed four pounds.

In the spring of 1779, contrary to my expectation, this thistle again made its appearance on and about the spot where the small piece was originally planted. There were between fifty and sixty young heads, which must have sprung from those roots, which had penetrated deeper than the gardener was aware of, although he was particularly careful in extracting them.

From these experiments it appears deducible, that no plants are more easily destroyed than the generality of thistles, or with more difficulty than this one; there being no soil, however poor, in which it will not vegetate, nor earth so stiff but it will penetrate; in proportion, however, as the soil is rich, will be its increase.

It were much to be wished, that an investigation of this evil had afforded a remedy: at present, none appears. It is, therefore, to be feared, that spudding or cutting them down close to the ground, once or twice in the spring, is the only operation the farmer can perform to prevent their bad effects in destroying his crops on arable land, and rendering his pastures unseemly.

As nature in the preservation of this plant seems to have exerted her greatest powers, it is possible that, in some future period, uses may be discovered to which it has not yet been applied.

To the ass it is the highest treat; and I have been credibly informed, that in some parts of Scotland, it is cut down as food for horses.

It would be well, if a plant so noxious in some respects could be rendered beneficial in others.

I am, &c.

W<small>M</small>. CURTIS.

A<small>RTICLE</small>

Article XXIII.

On a Disease the Norfolk Stock Lambs are liable to from eating self-sown Barley in Autumn.

[By a Norfolk Farmer.]

Gentlemen,

IN our open field-lands, where sheep are mostly kept, there are every summer large tracts of barley, in which, after harvest is ended, and particularly when it proves a wet Michaelmas, a great quantity of self-sown barley comes up in October. As the sheep then run at large, they are very fond of this crop, but it often proves fatal to them.

The cold dews in the latter end of October, and the fogs in November, generally hang in drops on the blade of this self-sown crop, longer than on the rest of the herbage; and the plants themselves, being of a cold watery quality, are thereby rendered still more unwholsome. After feeding on it for about a month or six weeks, the stock-lambs grow dull and heavy, rub themselves more than usual, lose their appetite, and waste till they die.

After they are once visibly affected, removing seldom saves them. The flesh of such as are killed appears loose and flabby. The viscera is of a livid hue, and very watery; the liver is greenish, and full of small knots or kernels; the blood is viscid, with a watery sediment, and very soon turns putrid. As soon as the symptoms appear, they are blooded below the eye: this, in the first stage of the disorder, is sometimes effectual; but no other remedy has yet been discovered.

The disease does not appear to be infectious; nor are either the ewes or wethers much subject to it.

I wish the Bath Society may be able to point out a remedy;—and am, &c.

Article XXIV.

Observations on the Mnyum Moss.

Gentlemen,

IF the following hint is worthy the notice of your Society, it is much at their service.

Wherever the mnyum moss, the red root, and the marsh pennywort, grow, there the water is uncommonly

uncommonly cold, and perhaps of a poisonous or mineral tinge. Grazing all low lands where such plants grow, as above-described, will occasion the death of many sheep, and cause some disorders in larger cattle.

I am, &c.

W. B.

ARTICLE XXV.

On the superior Quality of Grain produced from SET *Wheat, to that sown Broadcast.*

GENTLEMEN,

IN answer to your query, "Whether the *flour* of *set* wheat is *whiter*, or in any other respect preferable to that which is sown broad-cast?" I know of no comparative difference but what I conceive must necessarily arise from *fuller*, more *equal*, and *perfect corn*, than what is generally produced in the ordinary or broadcast mode: and as these circumstances are, I believe, always attendant on *set* wheat, there is less necessity of *throwing* and *dressing*, as practised by the farmer, or of *screening*, as practised by the miller; in course, the short answer will

will be, that *set* wheat will produce equally good flour, without the waste and trouble that attend the other.

On lands where wheat is *set*, the crop is free from those dwinged [shrivelled] diminutive grains that are so commonly found in even the best crops of that sown broadcast: of course, it is specifically heavier; and the proportion of flour exceeds the comparative difference of weight. Our farmers know this; they expect, and we* give a price exceeding that which is commonly given for the reputed best wheats that were sown broadcast.

I find the Gentlemen of the Bath Society are acquainted with our mode of setting wheat. Every farmer of industry and activity, who adopts this mode, will find the performance easy beyond his first apprehension; and feel himself repaid with an equal or greater quantity, intrinsically superior to what arises from the common practice. But I think, from the carelessness of droppers, three pecks of seed ought to be allowed to an acre.

Before I conclude, it may not be improper to inform you of an experiment I am now making in

* This Gentleman is a Farmer, a Miller, and a Baker.

planting

planting horse-beans. I had a piece of meadow-land, or rather wet waste, almost constantly under water in the winter, and not unfrequently so in the summer. It produced little but rushes, and (in the agricultural sense) other unprofitable aquatic plants. In February last, I opened, drained, and surrounded it with a ditch, the earth of which I threw into the hollow parts to render it level. It was then ploughed in the same manner as we plough the clover-lays for wheat, by turning the turf downwards. On this I set horse-beans, and the promise of a crop is abundant; there being from eighteen to forty-eight pods generally on a plant; and the rushes seem to be eradicated.

In order to bring a considerable hollow in this field to a level, I had the surface pared off a piece of furze ground pasture, and laid upon it. On the spot so filled up, I expected the best crop; but, on the contrary, here the blossoms dropt off considerably, the plants had a blasted appearance, and the produce was inconsiderable.

I am, &c.

Norwich, Aug. 15, 1779.

ARTICLE

Article XXVI.

An Account of the Cultivation of Siberian Barley.

[Sent to the Norfolk Society by the *Rev. Mr. Howman*, of Bracon in Norfolk; and transmitted to the Society at Bath by *Thomas Beevor*, esq; of Hethel-Hall.]

Gentlemen,

IN consequence of your having invited the members of the Society to communicate their experiments in any branch of husbandry, I am excited to relate the result of one which I made last year, in order to form some judgment of the advantage of cultivating Siberian barley.

A small inclosure, containing 3 acres, 1 rood, 2 perches, which had been under turnips the two preceding years, was sown with common barley, excepting one ridge of land in the middle of it, containing 20 perches, which was sown with Siberian barley the same day. The soil was very dry, and much inclined to a gravel.

The time of sowing was the 28th of April 1774, and the state of the experiment as follows:

Common

	Seed per Acre. Bushels.	Produce per Acre. Bushels.	Weight per Bushel. Pounds.	Total weight per Acre. Pounds.
Common Barley,	3	36½	52	1898
Siberian Barley,	3	32	58	1856

Excess of measure in favour of the common barley, 4½ bushels.
Excess of weight in favour of the common barley, 42 pounds.

A few particularities attended the growth of the Siberian barley, viz.

1. During the first three weeks after the corn came up, the Siberian was of a much deeper green, and had a much broader blade, than the common barley; after that time the difference gradually diminished.

2. The Siberian was in all its stages a fortnight forwarder than the common barley. It was mowed and housed accordingly.

3. The ears of the Siberian were much shorter than those of the common barley; being only from five to nine grains in length; whereas the ears of the common barley were from nine to thirteen.

From the first particular I had raised my expectations high in regard to the Siberian barley, and was consequently much disappointed at the appearance

pearance of the third. I then thought that the produce would be greatly deficient; but the size of the grains in a good measure prevented it.

The conclusion which I am tempted to draw from these two circumstances is this, that the Siberian requires richer land than the common barley. In my land, there appeared to be sufficient strength to produce all that luxuriance of growth which seems natural to the plant while in the grass, but not sufficient to support it in forming the ear. I am the more inclined to think this, having seen ears of Siberian barley of seventeen grains in length, which is the greatest length I remember to have observed in the common barley.

It may be remarked, that the circumstance of its requiring richer land does not seem to recommend it particularly to the county of Norfolk. On the contrary, the circumstance of its being forwarder than the other greatly recommends it to that county; for it seems evident from thence, that the Siberian barley may be, and perhaps ought to be, sown a fortnight later than the common barley.

A very large portion of our barley is constantly much damaged, both as to produce in measure and

and weight, by being sown too late,* in consequence of the necessity we are under of preserving some of our turnips as long as possible.

I am sufficiently aware that this experiment is not decisive; and that a single experiment, however decisive it may seem, is not properly conclusive; but I hope you will soon receive many others, and this may then contribute its mite towards forming an average, from which a just conclusion may be drawn.

I am, Gentlemen,

Your obedient humble servant,

Bracon near Norwich,   E. HOWMAN.
Feb. 11, 1775.

P. S. As it has been demonstrated before the House of Commons, that the weight of the flour of heavy wheat exceeds the weight of the flour of light wheat *more* than the difference between the respective weights of the grain; it may be safely concluded, that the same thing holds true with respect to heavy and light barley of the same kind: and with respect to the Siberian barley, it may certainly be concluded, that the weight of its flour

---

* The Norfolk Farmers seldom begin to sow Barley till May.

exceeds

exceeds the weight of the flour of common barley in a still more eminent degree; becaufe a part of the weight of the common barley arifes from a hufk, whereas the hufk of the Siberian barley is left upon the ear when threfhed. So that in this experiment, as the weight of the grain of the Siberian fo far exceeded the weight of the grain of the common barley, as *almoft* to compenfate for the great deficiency of meafure per acre, the weight of the flour of the Siberian barley per acre would probably have been found equal to, if it had not *exceeded*, the weight of the flour of the common barley.

As a bread-corn indeed in this county, barley feems to be out of the queftion; but the nourifhment muft be in proportion to the weight of the flour, however ufed.

Two things, however, want to be afcertained by well-authenticated experiments, viz. the quality of the Siberian barley in malting; and the quality of the beer made of that malt.

ARTICLE

Article XXVII.

On the Use of Fern-Ashes as a Manure for Wheat Land.

Aug. 10, 1779.

Gentlemen,

FOR several years past I have entertained a notion, that fern being burnt upon a fallow-ground would produce an excellent dressing for turnips and wheat; but have had no opportunity of making use of it myself, nor could I prevail till lately on either of the numerous farmers to whom I have mentioned it, to give it one trial.

The ashes of fern are stronger than any other, and must consequently, on account of the great quantity of salts it contains, be of infinite service in promoting vegetation.

I have great reason to believe, that fire has a beneficial effect on land, by reducing many parts of it to a more proper state for distributing its nutritive particles to the roots of plants. It must also destroy a great part of the roots and seeds of all kinds of weeds which may be in the ground, and consequently, in that respect, of very essential service.

In the courſe of laſt ſummer, (1778) a farmer who lived in my neighbourhood had a field of five acres under a fallow for wheat. It lay adjoining to a common which produces abundance of fern, and I obtained a promiſe from him to dreſs a part of it therewith. One other part of the field was dreſſed with dung alone; and the remainder with a mixture of lime and old mud taken out of a mill-pond at the bottom of the field. This laſt-mentioned dreſſing was well mixed, and laid on in a large quantity. No exact account was kept of the expence of the fern-dreſſing, nor of the quantity uſed;* we can, therefore, only gueſs at theſe particulars. It is, however, an undoubted fact, that 3s. 6d. is more than ſufficient to pay for cutting, drying, and carrying a waggon load in ſuch a convenient ſituation; and upon due enquiry, I was informed that about four waggon loads were laid on an acre; conſequently it muſt have been as cheap a dreſſing as could poſſibly be laid on it.

This field of wheat was reaped the 3d inſtant, as far as the fern was burnt, (which was two acres)

---

* Theſe are ſome of the omiſſions which render the experiments made by common farmers ſo indeciſive. They ought to be particularly accurate in eſtimating the quantity of manure, and the expence per acre; and in deſcribing the nature of the ſoil. Without this, a juſt concluſion cannot be formed on the ſucceſs of any experiment.

the

the wheat was in every respect the best in the field, being taller, stronger, thicker, the ears larger and finer, and the crop very clean from grafs and weeds. The reapers all declared they had not cut any wheat so good this season.

The part dressed with lime and mud was better than that dressed with dung only, that being the worst of all. The owner of the wheat and myself were both of the opinion, that the difference in respect of quantity of sheaves was in favour of the fern-dressed part, nearly as seven to five; but the difference with respect to the quantity of clean corn must certainly be in a greater proportion, by reason of the ears being so much larger and finer.

I hope the above relation (the truth whereof may be depended on) will be sufficient to convince those who are properly situated for carrying the experiment into further execution. But here permit me to observe, that successful as the above experiment proved, it was not fairly tried, for the following reasons: First, the fallow was not good for want of one or two more ploughings at the proper season; secondly, the fern was not cut until the latter end of August, and some in September, consequently

there

there could not be so much virtue in the ashes as there would have been in June or July.

I would recommend those who are disposed to try this dressing, to burn the fern at two different times—the first after the fallow has had the second ploughing—the other part after the ashes of the former are spread and ploughed in; by which means they will be more intimately mixed with the soil. I would also recommend that five waggon-loads of fern be burnt on an acre.

I am, your's, &c.

THOMAS PAVIER.

West-Moncton near Taunton.

ARTICLE XXVIII.

On the Cultivation of Heathy Ground.

[By a Suffolk Farmer.]

GENTLEMEN,

SOME years since I broke up ten acres of heathy ground, which had long been only a sheep-walk, and produced little else than furze, ling,[*] and mole-hill thyme.

[*] Erica, or Heath.

The soil was a loose blackish sandy gravel, and in general very dry. In March I turned it over with a whelming plough, about ten inches deep; and ran a pair of heavy harrows over it to get out the roots of the furze, ling, and other trumpery with which it had been overrun. These I burnt, and spread the ashes. In May, I ploughed it across with the same plough, harrowed, burnt the weeds, &c. and spread the ashes as before. In July, I ploughed it again, and spread thirty bushels of lime per acre. In September, I ploughed it a fourth time, with a common plough, harrowed it, and burnt the rubbish. By this time it was in fine tilth. In October, I sowed half of it with wheat, and the other half with rye; the former nine pecks, and the latter three bushels per acre; the winter proving favourable, the rye came up thick and grew winter-proud. In January, I ran a hurdle fence across, and turned in my sheep to feed it off—they remained there till the end of February, and left the field almost bare. I then top-dressed it with six bushels of lime per acre, which made the crop spring vigorously. It was as fine as ever I saw, and yielded me near five quarters per acre.

The wheat did not seem so strong as I expected; but toward spring it thickened, and I had near four quarters per acre. Both the crops were clean, and anfwered very well.

The next year I broke up fix acres more of the fame kind of land, and treated it in the manner defcribed above, till June, when I fowed it with turnips. They came up very well, and efcaped the fly. I gave them two good hoeings, leaving the plants fifteen inches fquare. The crop proved very good in quality, but the turnips were rather fmall.

At Michaelmas I turned in beafts and fheep, and in fix weeks fed them off. The land was very clean, and the manure left by the cattle had fo enriched it, that I thought there would be no great rifque in fowing it with wheat.

I gave it a good ploughing—fowed the wheat under furrow, and harrowed it down. It came up well, and the crop turned out near four quarters per acre.

In the autumn I laid on twelve loads of clay per acre; and in January, after a froft had made the clay break and mix eafily with the foil, I ploughed it in with the ftubble. In March, I ploughed it acrofs. In April, I ploughed it a third time,

time, and harrowed it fine: Then I sowed it with Zealand barley, ten pecks per acre; two weeks after sowing the barley, I threw in three pounds of Dutch clover. Both the clover and the barley soon made a fine appearance; the latter yielded five quarters per acre, and the former was a good thick plant at Michaelmas.

In the spring following I dressed it with forty bushels of soaper's ashes per acre, and in the summer mowed it twice: the first cutting was upwards of two tons, and the latter about twenty-five hundred per acre. I then broke it up for wheat, and had an excellent crop the summer following.

The lime, clay, and ashes, had doubtless their share in producing these crops; but I attribute my success chiefly to the repeated ploughings and burning the rubbish.

This land, which when I broke it up was not worth five shillings, is now worth twenty-five shillings per acre.

I am, &c.

G. L.

June 20, 1779.

Article XXIX.

Instructions for the Prevention and Cure of the Epizooty, *or* Contagious Distemper *among Horned Cattle.*

[Translated from the French of Monf. De Saive, Apothecary to the Prince Bishop of Liege, by Mr. Moreau, of Bath.

FARMERS have no need to be informed, how important a matter the preservation of their cattle is. The considerable advantages they reap from them when free from accidents, and the losses they suffer when distempers spread among their herds, are sufficient motives to make them feel the interest they have in preserving their cow-houses, stables, &c. from infection, and in using all possible means to prevent its progress. But as fatal experience has proved that the use of medicines, with the powers of which they were not well acquainted, has been frequently more prejudicial than salutary in the Epizooty; and that country people, by placing an unlimited confidence in pretended specifics, purchased at a very high price, have very often been drawn into a double loss, by the death of their cattle, as well as the expence of such drugs; it is thought the communication of an efficacious and cheap manner of treating cattle

cattle when attacked by this diftemper, and of the means to prevent their being fo, will be rendering an effential fervice to the public.

The moment any fymptoms of the diftemper are perceived, about a pint and a half of blood fhould be immediately taken from the beaft, except he has been ill a day or two, in which cafe he muft not be let blood; but in both cafes let the following draught be given:—

No. I. An ounce of the beft theriaca (Venice treacle) diffolved in a pint of vinegar, after which the back bone and the whole hide muft be well rubbed with a dry hair-cloth, to heat the hide and promote perfpiration. No drink fhould be given him but a white drink compofed of

No. 2. a handful or two of rye-meal in a pailful of clear water; and, fhould the beaft feem to want food, mix up fome crumbs of rye-bread with fome of the faid white drink, and give it him. The animal's mouth muft be wafhed twice a day with a cloth dipped in a mixture of

No. 3. Vinegar and water, (equal quantity of each) with a fpoonful of honey to a pint of it.

If on the second day the beast has not dunged, a clyster composed of

No. 4. A pint of water in which bran has been boiled, two spoonfuls of salt, and a small glass of vinegar, must be given and repeated every day 'till the evacuations are natural and regular.

Besides the above remedies, the following cordial mixture:

No. 5. A pint of clear water, the same quantity of vinegar, four spoonfuls of honey or syrup, and two glasses of brandy,—must be given four times a day, to facilitate and keep up perspiration; taking particular care to repeat the friction as directed above.

Should the beast still continue low and heavy, the draught No. 1. must be repeated, unless he should be found to be hot and thirsty, in which case use only the drink No. 2. On the fourth day, if he seems more lively and free from heat, purge him with

No. 6. Two ounces of salts, and one ounce of common salt, dissolved in a pint of lukewarm water, with two spoonfuls of honey. If this does

not

not procure four or five evacuations, repeat the clyster the same day.

This mode of treatment must be continued without intermission 'till the beast begins to eat; then you must only give him the white drink No. 2, and a little good fodder; or, some rye-bread dipped in stale beer, moderately sweetened with honey or syrup.

The exterior treatment consists in the application of setons in the beginning of the distemper, at the bottom of the dewlap, and of cauteries towards the horns, between which some weight must be fixed, such as, a stone of a pound weight, or more, wrapt up in a cloth, to keep it steady. This is necessary to keep the head warm. But above all, the friction must be closely attended to, in order to determine the critical efforts of nature.

It would be well also to evaporate vinegar in the cow-house, &c. and if it could be done without risque, blowing of a few grains of gunpowder twice a day in them, would be a very useful fumigation.

If, notwithstanding these aids, the beasts be not perfectly cured in ten or twelve days, they must
be

be continued without bleeding, unlefs the inflammation be very confiderable; but if, after all, the diftemper does not give way, the beaft muft be killed, and then too much care cannot be taken to bury it very deep, cover it over with the earth which came out of the hole, and a turf over all, in order to prevent the putrid vapours, which exhale from fuch carrion, corrupting the air, and fpreading the infection.

As to the prefervatives from infection, the principal, after having taken every precaution poffible to prevent its communication from other herds, confifts in wafhing the racks, troughs, &c. and the hide of the beaft every day, with plenty of water; and, as the generality of people feem to place great confidence in ftrong aromatic fumigations, they are advifed, inftead of the expenfive drugs of which fuch fumigations are compofed, to ufe fires made with the branches of green wood, throwing pitch on it to quicken the flames and perfume the air; thefe fires muft be lighted at fome diftance from the houfes for fear of accidents.

Common falt, given in fmall quantities every day to horned cattle, is reckoned an excellent prefervative, particularly in a learned differtation on

the

the contagious distempers among horned cattle, by Monf. De LIMBORG, M. D. and F. R. S. of London. It should be observed, that though the report of an Epizooty is often spread, yet all the disorders to which cattle are liable should not be attributed to this epidemical distemper, since they are not exempted from this even when not affected with any contagious distemper. Therefore, when a beast is taken ill, enquiry should be made if the infection be in the neighbourhood, as in such case, a suspicion of its being the Epizooty would be well grounded, and immediate recourse should be had to the remedies above-mentioned.

But as it often happens that cattle fall sick after having eaten bad fodder, or having grazed in frosty weather on the tops of herbs, &c. when covered with ice and snow, (to prevent their doing which, all possible care should be taken) to these accidents only are frequently to be attributed the sickness and death of many beasts which fall victims to them.

There is another accident no less dangerous, to which cattle are liable, which is, the washing them with waters prepared with different sorts of poisons, especially with arsenick to kill vermin; these

waters

waters occasion an itching in the skin, which obliges the animal to lick himself; in doing which he sucks in the poison. It is evident that such pernicious practices may occasion as fatal disasters and unhappy losses to farmers, as even the Epizooty itself; it cannot, therefore, be too much recommended to them, to forbear the use of such things, which never fail doing the mischief above described.

Article XXX.

On the Construction and Use of Machines for Floating Pastures, and for Draining Wet Lands.

[By a Gentleman in Wiltshire.]

Gentlemen,

I Beg leave to propose to your consideration, the propriety of offering a premium for the most simply constructed Machine, which shall fully answer the purpose of raising water either for draining or floating land; such Machine to be worked by a small running stream, where there happens to be a sufficient fall of water:—And another premium for a Machine capable of being worked by wind to answer the same purposes.

Machines

Machines of these kinds, properly constructed, would at once improve lands, enhance their value, and reward the ingenuity of persons skilled in mechanics.

There are many instances of pasture lands being raised two-thirds in their annual value, by being floated; viz. from one to three pounds per acre.

Any gentleman, therefore, who has twenty acres of land capable of such improvement, might well afford to expend fifty pounds in erecting such machines as would effectually answer the purpose; especially as land so watered affords, in spring, the finest pasturage for ewes and lambs.

Where the wind is to be the moving power, the chief merit of the engine will consist in its being so constructed as to be worked with the least attendance, and turn about easily of itself to receive the wind from every direction. There is a certain point between the small self-working wind machines, and the large ones which require constant attendance; which point, could it be exactly hit upon, would determine what is the largest size of the *vanes*, and what the greatest weight of the machinery, that can be constructed so as to be useful without attendance.

An

An attention to the construction of such machines as that erected on the top of Newgate, to work the ventilators, might throw some light on this subject. This turned on a pivot, to receive the wind from every quarter, by the impulse of the shifting winds impressed on the horizontal vane; and its force was equal to what would be required to raise a considerable quantity of water to the height of two or three feet, which is generally sufficient for the purposes of draining or floating land.

It seems to me that a method might easily be found to check or counteract the force of the wind when too violent, and even to stop the motion of the machine, by means of any great increase of the wind's force.

I therefore doubt not, but if the exercise of ingenuity were called forth, by the offer of suitable rewards, it would be exerted successfully in constructing some engine which would effectually answer the purpose.

I am, &c.

Marlborough, Dec. 8, 1779.

ARTICLE

Article XXXI.

Experiments to ascertain the Use of Soaper's-Ashes and Feathers, as Manures.

[Communicated to the Society by an Essex Farmer.]

Gentlemen,

IN April last I top-dressed half a field of clover with ashlep, or soap-boiler's ashes, at the rate of sixty bushels per acre, leaving the other half in its former state. The effect exceeded my expectations. There was no apparent difference either in the soil, or in the crop it produced last year; but in consequence of this manure, the dressed part of the field produced, the last summer, nearly double the quantity of hay as the other.

I have also used this manure with great success in cold wet spungy meadow-land. It has apparently dried it, and, by its warm quality and the salts it contains, made it produce much greater crops of grass than before. I would therefore recommend it for both the above-mentioned purposes, if it is not already in use in your country, and can safely answer for its success. The farmers here will readily give from twelve shillings to a

guinea

guinea for a waggon-load, and fetch it five or six miles, and they find their account in so doing.

Another experiment, I think, may not be improper to mention: In October last, having a field ready for sowing wheat, I manured one acre of it with ten bushels of old feathers, procured from an upholsterer, ploughing them in as they were spread; and from the success of this experiment, am persuaded there is no kind of manure for either wheat or summer corn land equal to it. The acre thus manured produced me near forty-eight bushels—the other part of the field not twenty-eight bushels on an average. The quality of the land, and of the seed sown, was equal; the extra produce could, therefore, only proceed from the manure.

I wish all possible success to your undertaking; and am,

    Your's, &c.

          J. B.

Oct. 4, 1777.

ARTICLE

Article XXXII.

On Planting Boggy Soils with Ash; and the Slopes of steep Hills with Forest Trees.

[By Mr. *Fletcher*, near Northleach, in Glocestershire.]

' Gentlemen,

HAVING been pretty largely concerned in planting Forest Trees, on various soils, for more than twenty years, and tried different methods, I have found, by repeated experience, that no land whatever is so proper for the growth of *Ash* in particular, as swampy, rushy, and boggy soils. I have planted Ash on land which was so boggy and rotten, that the men were obliged to stand on boards, to prevent their being mired, and which, from its situation, could never be drained so as to render it fit for the cultivation of corn or grasses. It was astonishing to see their growth. Facts can be easily produced to prove, that such land (not worth a shilling per acre for any other purpose) has in divers places produced, in thirteen or fourteen years, from sixty to seventy pounds worth of the finest ash poles, at a moderate price, besides leaving a proper quantity of oaks, &c. sown with them, for maiden timber. Where labour is not very dear, an acre of such land may be
planted

planted with four thousand five hundred Ash sets, (which are a proper number) for eleven or twelve pounds. This, Gentlemen, I have frequently done, and I think it is an object worthy the attention of your Society.

A small expence of weeding, pruning, &c. will arise for two or three years after planting, but then it is over; and I think no method of cultivation can possibly prove so advantageous on boggy soils.

A dead foxy soil, or land overrun with furze and fern, will also answer exceedingly well for copsing; and, as wood is in many places become scarce, I think planting ought to receive every possible encouragement.

There is another kind of land, which, although fit for scarcely any thing else, I have planted with great success; and that is, the sides of very steep hills, particularly in a northern aspect. If there is any depth of soil, Ash-trees will do very well there; and for Scotch firs, and beech, it is a very proper situation.

I am, Gentlemen, yours, &c.

JOHN FLETCHER.

Oct. 17, 1777.

[We are too apt only to take the better side of a matter in which our judgment or our interest is concerned. An acre of Ash, &c. may be planted for eleven or twelve pounds; but a very material expence is here omitted, the fencing them from the incroachments of common cattle, &c. which cannot amount to less than eight pounds per acre. We do not insert this to prevent planting, but only that others may not be misled.]

Article XXXIII.

Mode of Cultivating Turnips in Suffolk.

[By a Gentleman Farmer near Ipswich.]

Gentlemen,

AS the Turnip husbandry, properly managed, is the foundation of the best system of Agriculture hitherto discovered, I take the liberty of sending you an account of our method of raising that valuable root, which we find very advantageous, both as food for cattle, and as a cleansing fallow for succeeding crops of grain.

In November, we plough in our wheat stubbles, and give the land four ploughings afterwards, at such times as suit our convenience. Previous to the last ploughing, which should be in the latter end of June, we cart on twenty loads (as much as three horses can draw) of rotten dung, or muck, from the farm-yard, turned up in April, or early in May. Sometimes, as lime is the readiest and cheapest manure in these parts, it is used instead of dung, but I think the latter preferable.

One quart of seed is sufficient for an acre. New seed will come up three days sooner than old.— What is ploughed for the last earth should be sown the same day, else, unless rain falls, the ground will be too dry for the seed to vegetate. When the turnips are within three days of being fit for hoeing, if the weather be dry, we run a pair of light harrows over the field, in a direction contrary to that of sowing, and before they are hoed the first time. We find this to be of much service.

The Turnips should be hoed with a seven-inch hoe, and left full twelve inches* distant from plant to plant. We always hoe them twice, and by that means gain near double the weight in produce.— The labourers, who by use become very expert in this business, have three shillings and six-pence per acre for the first hoeing, and two shillings and six-pence for the second, with small beer.

* Fourteen or even sixteen is better.

Crops

Crops vary according to the quality of the land, from forty shillings to three pounds per acre in value,* and are mostly fed off in the field.

In feeding them off we generally first draw off a rod in width round the field. This is done to prevent the cattle from spoiling them, by getting near the hedges for shelter in bad weather. The farmer first puts in his beasts†—then his best wether sheep, and lastly his lambs, which eat up all the refuse left by the others.

As soon as the field is cleared of its stock, we plough it up for barley, and give four earths.—Sow three bushels of barley per acre, half above and half under furrow. Fourteen pounds of red clover seed‡ is harrowed in with the barley, and the land rolled after the barley is come up. The

---

* This is a very indeterminate quantity, as what may be worth forty shillings in one place may be worth three pounds, or more, or less, in another.

† We cannot agree with the custom of turning in beasts before sheep intended to be fatted. If the beasts are intended to be fatted, we apprehend the intention will be very materially lessened, by their running about. A stalled Ox ever while you live!—except in fine meadows.

‡ Too much. Eight pounds are enough, if the seed be good; but on light lands, five pounds of black grass, or hop clover milled, and five pounds of broad clover, will answer most incomparably well.

produce from thirty to forty bushels per acre. An excellent crop of clover generally succeeds the following season, which, after being once cut for hay and then seeded, is ploughed in for wheat.

The Turnip is our fallow, and the better that crop is, the finer are the crops of barley and wheat which succeed it. When the land is too wet to feed the Turnips off, we draw and carry them on some contiguous pasture.

I am, your's, &c.

N. T.

Nov. 19, 1777.

Article XXXIV.
On raising Potatoes from Seed.

[By the Rev. Mr. Lamport, of Honiton.]

Gentlemen,

I TAKE the liberty of recommending to your consideration, to offer a premium for raising Potatoes from seed; and also of sending you the method of raising them, prescribed by the ingenious Dr. Hunter, which, from several trials, I have

have found to anfwer all the encomiums that gentleman has beftowed on it.

My reafons for making this propofal are,

1*ft*. It perfectly coincides with a material part of your benevolent plan to increafe the quantity of food for the poor: As an acre cannot be planted in the common method for lefs than twenty-four or twenty-five fhillings, the mere purchafe of the roots; the procuring plants from *feed* will not coft more than five fhillings:* So that Potatoes of the beft kinds, to the amount of twenty fhillings an acre, will be annually preferved for food, inftead of being

* The principal advantage arifing from raifing Potatoes from feed, will be the obtaining a better or larger fpecies, which will be more profitable in its cultivation for fome years, than a fpecies planted for many years together, without change: But furely no immediate advantage can happen to the planter! Let the oppofed accounts afcertain the fact.

	£.	s.	d.		£.	s.	d.
Sowing the feed	0	5	0	Potatoes —	1	4	0
Rent of the land	0	15	0	Rent ——	0	15	0
Hoeing ——	0	9	0	Hoeing —	0	9	0
Planting ——	0	8	0	Planting -	0	8	0
	£.1	17	0		£.2	16	0
Difference	0	19	0				
	£.2	16	0				

Produce of fown 150 bufhels, at 2s. —— £.15 0 0
Produce of the planted 200 bufhels, at 2s. — 20 0 0

cut for planting. This will be a saving to the four counties, in proportion to the number of acres planted in the method recommended.

2*dly*. Various sorts of Potatoes are produced from the seeds of the same apple, and by this means *new* kinds would be introduced, some of which might be more valuable than any yet raised.—The farmer will have it in his power, at his option, to preserve for his own use the best kinds, or those best adapted to his soil, and to dispose of the rest either by sale, or as food for his cattle.

3*dly*. Potatoes will, after a few years, unavoidably degenerate, and decrease in produce; while those newly raised from the seed will produce, at least, one third more than those which have been usually propagated in the county, or which can be procured from other counties, unless the farmer could certainly know that the Potatoes he purchases were newly raised from seed also.

For these reasons, this method appears likely to prove of great publick utility, and to deserve a premium for encouraging it, especially as the process is neither difficult nor expensive.——The method is as follows:

Let

Let the farmer, or gardener, gather the apples of his potatoes in October, and hang them up in a warm room till Christmas; then wash out the seeds, spread and dry them on paper, and preserve them from damps till the spring. In March sow them in rows one foot asunder, in ground well prepared; and when the plants are three inches high, gently earth them up. About three weeks after, transplant them in land well dunged and made fine, and dig them up in autumn.

By this means you may have great varieties, and your crop next year will be large both in size and quantity.

Article XXXV.

On the Method and great Advantages of extracting the Essence of Oak-Bark for Tanning.

[Communicated to the Society, Dec. 7, 1777.]

Gentlemen,

AS there are large quantities of oak-bark annually imported into this kingdom, the bulk of which renders the freight very high, and consequently makes the article very dear, besides endangering

gering its being spoiled by getting wet, it would be of great advantage to the community if the astringent qualities of the bark could be extracted on the spot where it grows, and reduced to the consistency of a thick essence. By this means, the virtues of a large bulk of bark might be collected into a small space, which would make a great saving both in the freight and inland carriage, and render it a staple commodity for trade.

On considering the subject attentively, I am of opinion, that the scheme is practicable, and would answer extremely well. With respect to the process, this extract must first be made either by decoction or infusion; and then the watery particles must be evaporated, to reduce it to the consistency desired, in such a manner as not to lose any of the qualities necessary in tanning.

Suppose the operator has at his command a common family brew-house, with its necessary utensils: let him procure a ton of good oak-bark, ground as usual for the pit; and having placed a strainer to the mash-tub; fill it two-thirds with the bark; heat as much water, nearly boiling, as will sufficiently moisten it, and mash it well together. After it has stood about two hours, draw it off clear,

clear, and put it into a cask by itself. Make a second extract with a smaller quantity of boiling water than before, so as to draw off a quantity nearly equal to the first, and put that also into the same cask with the former.

These two extracts will probably contain in them as much of the virtues of the bark as the quantity of liquid will absorb.

A third extract, rather more in quantity than the other two, may be made from the same bark, and as soon as drawn off, should be returned into the copper again when empty, and applied for the first and second mash of a quantity of fresh bark, as the three extracts may be supposed to have carried off the virtues of the first. Then proceed as before till all the bark is steeped, and a strong liquid extract is drawn from it.

The bark, when taken out of the copper, may be spread in the sun to dry, and serve as fuel in the succeeding operations.

The next process is, to evaporate the watery particles from the extract, by a gentle heat, till it comes to the consistency of treacle. This may be done either by the air and heat of the sun, or by

the

the *fill*, or iron pan, over the fire. For this experiment, shallow vessels will be sufficient. It must not be heated to boil; for that would be likely to drive off by steam* what we want to retain. Let the evaporating vessel be covered, during the process, with a wooden lid, through which a number of holes are bored with a gimlet, as the steam will fly off much quicker this way, than if left uncovered; and for this reason, that in the latter case, the air, pressing on the whole surface, would prevent the steam from rising so freely as it will do through a number of small apertures.†

You will be pleased to observe, Gentlemen, that my first object was, to get this extract made in America, from whence large quantities of rough bark were annually imported: but the unhappy state of the colonies prevents its taking place there at present. In some future time, perhaps, the attempt might succeed; and as the sun is much hotter there in summer than with us, the evaporation might be made by its heat, without the expence of fire.

* In this point, we think our correspondent mistaken; being of the opinion, that boiling would not cause any evaporation of the essence itself.

† In this respect also we think the Author mistaken.

In

In the colder climates, such as Canada, where Dutch stoves are used five months in the year, the same fire would serve for evaporation; so that when the process, which is not difficult, becomes generally known, the country people might collect bark in the season; and during winter, when they have little to do, extract its essence. But were it carried on in a manufactory, the heat might be so frugally applied as to occasion little expence; for the evaporating vessels might be so constructed and placed, as for the steam to empty itself into the steeping tub, and there condense itself into hot water. This would save both time and expence.

The universal use of leather, and the great scarcity of oak bark, make these considerations of great importance to the publick; and it were much to be wished, that fair trials were made, both of extracting the essence, and tanning leather with it when extracted, with an accurate register of the expence attending each process.

I am, Gentlemen, &c.

Article XXXVI.

On Drilling Pease.

[By a Gentleman near Taunton.]

Gentlemen,

PERHAPS the following observations on a crop of pease, may not be wholly unworthy your notice.

A farmer in my neighbourhood sowed a few pease in drills, in a common wheat field, in the beginning of November last, for the use of his family. We had some meals of them well grown, when the price was two shillings and six-pence per peck; and when they were sold in Taunton market for sixteen-pence per peck, the ripest being gathered from four of these drills only, (from which none had been gathered before) produced two pecks of pease; and as the drills were only eighteen feet in length, and two feet distant from each other, the whole space of ground occupied by the four drills was no more than sixteen square yards: from whence it appears, that one acre of ground, statute measure, would have produced upwards of six hundred pecks of green pease at the first gathering; or, if you calculate by the acre

acre of fifteen feet to the perch, (which I take to be the faireft way) the produce would be five hundred pecks, which, at the then current price, amounts to thirty-three pounds fix fhillings and eight-pence. An ample encouragement for trying this method on a larger fcale!

I am, &c.
T. PAVIER.

[We are obliged to this as well as the reft of our correfpondents, for communicating accounts of experiments, and making calculations thereon; but muft beg leave to obferve, that the advantages arifing from any experiments made on a *fmall fcale*, will not be proportionably great, when that fcale is confiderably enlarged. In the inftance before us, we doubt not the truth of the account given: the calculation is right, and the profit obvious: but we cannot think that the produce of an acre (much lefs of a larger quantity) would fell for the fum mentioned. An additional quantity of five or fix hundred pecks of green peafe, at the time they are fold at fixteen-pence a peck, would immediately reduce the price in any country market. We by no means hint this with a view to difcourage experiments and calculations; but merely to guard againft expectations of profit too fanguine to be realized.

realized. If the above-mentioned crop were sold at only nine-pence a peck, the farmer would be well paid for his labour.]

Article XXXVII.

An Account of the Culture of Siberian Barley, in 1774.

[Transmitted by the Norfolk Agriculture Society.]

THE intention of the Norfolk Society being to ascertain the positive produce of the grain, and also the comparative produce in the two methods of sowing in broadcast, and of setting by the dibble, the following experiments were made:—

The ground chosen for this purpose was a sandy loam, containing, exclusive of hedges, 1 acre, 3 roods, 6 perches, 19 square yards, statute measure. It had been in grass three years at the autumn of 1772, when it was broken up. In the summer of 1773, it bore oats, and in autumn the same year had three ploughings; in the spring of 1774, it was ploughed three times more. Being then in good order, it was divided into two parts.

No. I.

No. I. contained 3 roods, 2 perches, 9 square yards. This was sown with Siberian Barley, by dibbling, from the 9th to the 13th of May 1774. The distance of the holes was about six inches one way, and four the other. The directions given were to drop not more than two grains into each hole, and they were in general executed exactly. The quantity of grain sown in this part was one bushel and a handful. On the 18th of June, the stalk was two feet in height, and the ear completely formed in the sheath. It proved a very wet summer, and the rain beat it down. It was cut September the 6th; some of it grew before it was carried home, and some could not be threshed out of the straw; but the produce was forty bushels one peck Winchester measure, each bushel weighing fifty-four pounds.

No. II. (the other part of the field) contained 1 acre, 4 perches, 10 square yards. This was sown under furrow with Siberian Barley, May the 10th and 11th, 1774, by which means this part had one ploughing more than the other. The quantity of grain sown was four bushels. This was sooner and more beaten down by the rain than the other, as it grew thicker; but as it was cut sooner, (viz. August 26) it was not so much damaged, the other having

having begun to grow before it was cut. This produced forty-nine bushels Winchester measure, each bushel weighing fifty-four pounds. Bread was made of this barley, mixed with wheat flour, in the two proportions of half each, and of two-thirds barley and one-third wheat. This was repeated several times, and the *latter* proportion was thought to make a sweet and pleasant bread.

Comparison.

BROADCAST.

Time of Sowing.	*Quantity of grain sown.*
May 10, 11.	4 bushels.
Time of Cutting.	*Produce.*
Aug. 26.	49 bushels, 54lb. per bush.

DIBBLED.

Time of sowing.	*Quantity of grain sown.*
May 9 to 13.	1 bushel.
Time of Cutting.	*Produce.*
Sept. 6.	54 bushels, 54lb. per bush.

So far as a single experiment can be conclusive, it appears,

1*st*. That on land in good order, the produce of Siberian barley is great.

2*dly*. That

2*dly*. That the method of setting by the dibble is most productive in the proportion of about one-tenth part.

It is to be observed, that three-fourths of the seed is also saved; but this is allowed for the extra expence of dibbling.——The increase on the dibbled part was forty from one yearly.

Article XXXVIII.

On a new Oil Manure.

[From a Gentleman Farmer in Norfolk to the Norfolk Society; and communicated to the Society at Bath by Thomas Beevor, esq; of Hethel-Hall.]

Gentlemen,

I Now take the liberty to lay before your respectable Society a Receipt to make a manure for the improvement of lands, which I have with much pains been so happy as to find out, and which bids the fairest of any thing yet thought of, for general benefit. It is equal to either muck [*dung*] or oil-cake, both of which are allowed by all who use them to be of great utility; but there are few who can find sufficient quantity of the former, and the latter is too expensive for general use.

The following is the composition of the Manure here recommended, with the expence, for one acre of land:—

Rape, or train-oil, 6 gallons, at 2s. 6d.	—	£.0 15 0
Sea-sand, 6 bushels, at 2d.	—	0 1 0
Coarse Salt, 2 bushels, at 1s.	—	0 2 0
Malt-coombs, 24 bushels, at 4½d.	—	0 9 0
		£.1 7 0

The method of preparing it is, to spread the coombs on the floor about four inches thick—then sprinkle the salt as level as you can; throw on half the quantity of sand, and half the quantity of oil, out of a watering-pot—turn it and rake it well—afterwards add the rest of the oil and sand as before—turn it well 'till thoroughly mixed, and then throw it in a heap for use.

As the prolifick quality oil-cake is only in proportion to the oil it contains, the composition I now recommend must be preferable, having a much greater quantity of oil in it; and as malt-coombs are a manure of themselves, especially for turnip-land, at about eighty or ninety bushels per acre, I dare venture to assert, that twenty-four bushels,
with

with the addition of oil, is equal to the above quantity, or even to twelve loads of muck. The sand, and salt mixed with it, not only occasion it to imbibe the oil more freely, but likewise give it a better body for the conveniency of spreading on the land.

Some Gentlemen may think that the quantity of salt is too little; but I am truly convinced of the contrary, having found by experience, that a ton, or even a ton and half, has not answered so well as three or four hundred. The case is very similar with regard to lands near the salt-marshes, where the tide sometimes overflows them; and it is well known by those who occupy such lands, that nothing will grow for three or four years; but afterwards they will become very fertile. The reason I shall not take upon me to give, but have found it so by my own experience.

If Gentlemen will make trial of this manure, I have no doubt but it will answer their utmost wishes. Some farmers here have used only half the above quantity per acre, notwithstanding which they had good crops.

J. C.

Wells, Norfolk, June 12, 1776.

P. S.

P. S. I was observing a few days since a field of barley belonging to Mr. Tuttle, of Wareham, that was overflowed by the tide two or three years ago; and nothing has grown upon it since till the present year; but there is now a prospect of the finest crop I ever saw, especially on that part of the land which was overflowed. This I consider as a proof that too much salt is very injurious; and would therefore advise every farmer who makes use of it, to adopt the quantity as may be found necessary according to the quality of the land.

[On this Gentleman's manure we beg leave to remark, that nothing appears to us against its becoming generally useful but the extra expence that must attend it, from the great price of oil, and the expence of the carriage of sea-sand and drofs salt, in most inland situations. Yet to those who live near the coast, and are willing to use oil at its present price, we have no doubt of its answering the purpose; perhaps the common sand may be as proper as that from the sea-shore; but in this case we think more salt will be requisite.]

ARTICLE

ARTICLE XXXIX.

An Account of a Mode of Weaning and Rearing Calves, by a Norfolk Farmer.

[Communicated by the Norfolk Agriculture Society.]

MR. Whitby, of Wallington, did, between the first of December 1776, and the first of April 1777, wean, and rear on his farm, ten cow-calves, and thirteen bull-calves, by the method following:—At three days old, they were taken from the cows, put into a shed, and fed with flet (skimm'd) milk for one month, allowing three quarts to each calf morning and evening. When a month old, they were fed with the like quantity of milk and water, morning and evening, with hay to feed on in the day-time; and at noon they were fed with oats and bran equally mixed, allowing half a peck to one dozen calves. At two months old, they were fed only in the morning with milk and water, they had hay to feed on in the day-time, and at evening, instead of noon, had the same quantity of bran and oats, with water to drink. They were fed in this manner until the middle of April, when they were turned out to grafs all day, and taken into a shed at evening; and fed with hay until there was plenty of grass, and the weather grew warm.

Such of the calves as were weaned in March were continued to be fed with milk and water every morning until Midsummer. All the said calves are in good health and condition; and the Society allowed the premium offered on that head the preceding year.

ARTICLE XL.

On raising a Crop of WHITE OATS *and* GRASS-SEEDS.

[By a Berkshire Farmer.]

GENTLEMEN,

AS I observe in your advertisements frequent invitations for the correspondence of practical farmers, the following account, being a just one, is much at your disposal.

In the year 1774, I bought twelve acres of land, which had been sown with white oats and grass-seeds to lay down for meadow.

On examining my plant of clover, &c. after the oats were off, the couch-grafs and clutter, [*weeds*] from its having been laid down very foul, had almoſt totally deſtroyed the young grafs; there being ſo little left, that no profit could be expected from letting it ſtand. I therefore had it ploughed up immediately; and my crop of oats having been houſed pretty early, I gave it a good tillage. After getting out as much of the couch-grafs and rubbiſh as I could, before the winter came on, I had it ploughed up in rough ſtetches, (or ſingle ridges) that it might have every advantage of the winter's froſt to mellow it, which it did very effectually. In the ſpring my plough went to work again: we found the roots of couch, &c. which had been diſturbed by the tillage in autumn, generally dead. Then I ſowed it again with white oats and grafs-ſeeds, not the rubbiſh of a hay-loft, which abounds generally with the ſeeds of numberleſs weeds; but the beſt I could collect. My neighbours perſuaded me to dung it; but this I omitted till the crop of oats was got in. I then dreſſed it well with the beſt ſtable-dung I could procure. My crop of oats was but indifferent; but my grafs the following ſummer, being of the moſt excellent kinds, was full two tons per acre at one cutting. I did not mow it a ſecond time, as I wanted the ſeed—nor did I cut

it

it the year following. At the latter end of the succeeding year I dressed it again, and have had as good crops since as from meadow-land held at one-third more annual rent.

I am, &c.
C. T.

Berks, Aug. 24, 1779.

ARTICLE XLI.

Answers to the Society's printed List of Queries.

[Communicated by EDWARD SAMPSON, esq; High Sheriff of the County of Glocester.]

Henbury, Dec. 14, 1778.

GENTLEMEN,

I HAVE the pleasure herewith to transmit answers to the list of Queries, with which you some time since honoured me. If they in the least degree answer the Society's expectations, it will be a satisfaction to

Your very humble servant,

EDWARD SAMPSON.

Answers

Answers to the Queries proposed by the Agriculture Society at Bath:

To the First Query.—Cone wheat, and blue ball, on strong clays, and deep rich loams; the several kinds of Lammas wheat on loams, sand, gravel, and stone-brash land. Barley most natural on sandy, gravelly, and stone-brash; but it will return large crops on clays, although the grains are more coarse and brown.

Pease for culinary uses on sands and loams;—for pigs, on clays, gravel, and stone-brash.

Beans on strong clay and deep loam, the same as cone wheat.

Vetches on gravelly soil and stone-brash.

Turnips on every kind of soil, with good and repeated ploughings, and proper manures:—most natural on a sandy loam.

Cabbages on strong deep clays and rich loams.

Carrots on deep loams abounding with sand, and not too stiff; and on any deep light soil duly cultivated.

The quantities of seed depend much on the season and time of sowing. Wheat from seven to ten pecks per acre. Barley from ten to sixteen pecks. Pease and beans ten pecks if drilled, twelve if planted, sixteen if sowed, and earthed or harrowed in. Vetches from eight to ten. Turnips from ten to twenty-four ounces. Much depends on the skill of the sower. Cabbages and carrots have the like dependance. The average produce cannot be ascertained with precision, because of blights, mildews, earth-grubs, and many other accidents to which all sorts of grain are incident; and, exclusive of these, much depends on the nature of the soil and mode of cultivation.

To the Second.—On clay and loamy soils, if old arable long in tilth, the following course is generally practised: 1. Turnips, as a fallow-crop; 2. Barley; 3. Clover, mowed early, and then fed; 4. Wheat, on one earth; 5. Pease, or beans; 6. Wheat, then Turnips. If a new farm from pasture, 1. Beans, or Pease; 2. Wheat; 3. Barley: *or*, 1. Turnips; 2. Barley; 3. Clover; 4. Wheat, and then Turnips again.

On light thin and stoney soils, 1. Turnips; 2. Barley; 3. Clover, mowed early and fed till Midsummer,

summer, then let it grow, and plough it in for wheat. Vetches in winter, and fed off for Turnips.

To the Third.—For stiff clays, sand in due quantity; for light sand, clay in due quantity; and for both, lime duly prepared. Lyas lime for light sands; marble lime for heavy soils.

For gravelly and loamy land, yard-dung, lime-chalk, and shovelling of highways, in composts.

For moorish and cold soils, gravel, highway-earth, very small stones, coal-ashes, soaper's-drains, and pig's-dung.

For cold wet lands, no manure effectual without draining, and then the same as for the last.

For stone-brash land, any kind of manure laid on in a half-rotten state. The quantity per acre must be learned from experience. It is better to lay on at twice than too much at once. The season from February to September. The time of lasting is according to the understratum, which, if compact and warm, will render the manure durable; if loose, or a cold clay, it will soon be gone.

To the Fourth.—No new discovery of manure in the south parts of Glocestershire, except about Bristol. The dung and urine of pigs, fatted by the wash of the distillers, are found to be excellent manure for any kind of land, but more especially cold clays. The lees or suds of soap-makers are also found of great use, as well as the urine of pigs, by being sprinkled over pastures in the same manner as the roads are watered about London. Care must be had to due quantity, or the verdure will be destroyed. Experience is the best guide.

To the Fifth.—All dressings on cold wet lands will be very ineffectual, unless the lands are first dried by under-draining. Soot is the most beneficial, only the hay will smell of it.

To the Sixth.—Stone is the best and most lasting; wood is a substitute, and will be lasting also if constantly wet; if not, it will soon be rotten, and then the trenches will close.

To the Seventh.—The wood which stands best against west winds, on high exposures, is the beech and the black mountain sallow, (*Salix Latifolia Rotunda,* being the thirteenth species of Miller) with a plumb-tree leaf; on moorish and boggy-ground, the black alder.

To the Eighth.—Lucerne is cultivated by very few, and those more for fancy than profit, as it will bear no rival, but must be kept hand-weeded, or it will soon decay; nor will it succeed even with such care on lands of a cold or moist understratum.

Sainfoin is cultivated on dry, gravelly, and stone-brash land, when the understratum is not of a close compact texture, but of a loose open stoney nature, or chalky. It answers well in the broadcast method. The cause of its often failing is owing to the nature of the land, rather than to the mode of cultivation.

Burnet (the *Pimpernella Sylvestris* of Ray, *Pimpernella Sanguforba major* of C. B. 160, and *Sanguforba* of Linnæus) grows naturally in moist clay meadows, in this county; but the cattle will prefer almost all other common plants found in those pastures to it. The lesser Burnet (or *Pimpernella Sanguforba minor hersuta* of C. B. B. and *Poterium* of Linnæus) delights in a gravelly dry soil, and is frequent in healthy sheep-pastures, and eaten greedily by those animals.

To the Ninth.—Turnips are generally sown as a fallow crop, after the land (of any sort) is well tilled, cleansed from weeds, and dressed with yard-dung, lime,

lime, or any compost. We generally sow them about Midsummer, and hoe them twice; they may be effectually preserved from the fly, if, as soon as the seed-leaf appears, wood-ashes be sown over them as often as it is washed off by dews or rains.

To the Tenth.—The drill is preferable to the broadcast method, in loose or loamy land; but not in clays or stoney soils.

To the Eleventh.—The comparative advantage of oxen is great where they are bred by the farmer who uses them, and fed on commons in summer, and on straw in winter, till three years old, (but not so much where they are bred in inclosed lands, or bought at four years old) and worked till six or seven; they are less liable to sickness than horses; and if accidents befal them, they are of some value. Two oxen will do more work than one horse of equal value with them, nearly in proportion as six to four, and they cost less in keep.

To the Twelfth.—In places subject to rot sheep, fold them before the dew falls, and keep them in fold till it exhales, in spring and summer; in winter attend to this as much as the weather will admit; and feed them in the fold, or on turning out, with

hay

hay on which salt has been sprinkled at stacking up at harvest. It is a known truth, that the pastures (though marshes) which are overflowed by the salt water at the vernal or autumnal high tides, never rot sheep, but are an antidote to the disease, if the infected are depastured thereon while the disease is recent.

To the Thirteenth.—Chiefly in the cross-tree, pot-hook-drail, swing-plough, which with two horses will plough most kinds, and with three horses any sort of land; having a point to the share for stoney lands, and no point in lands that are not stoney.

RURICOLA GLOCESTRIS.

ARTICLE XLII.

On the great Increase of Milk from feeding Milch-Cows with Sainfoin.

[By an Essex Farmer.]

GENTLEMEN,

IN looking over your list of premiums, I was much pleased to find your Society had encouraged the cultivation of Sainfoin. In this neighbourhood,

bourhood, we have many large fields of this excellent grafs, and find it the beft and moft profitable of any that we raife.

As the roots ftrike deep in our chalky foil, this plant is not liable to be fo much injured by drought as other graffes are, whofe fibres fhoot horizontally, and lie near the furface. The quantity of hay produced is greater and better in quality than any other. But there is one advantage attending this grafs, which renders it fuperior to any other; and that arifes from feeding it with milch-cows. The prodigious increafe of milk which it makes is aftonifhing, being nearly double that produced by any other green food. The milk is alfo better, and yields more cream than any other.

I give you this information from my own obfervation, confirmed by long experience; and if your farmers would make trial, they would find their account in it far more than they expect.

Your's, &c.

J. B.

Near Saffron-Walden, Feb. 1778.

ARTICLE

Article XLIII.

An Account of the Success attending the planting Moor-Land with Ash-Trees.

Gentlemen,

BEING lately in the county of Essex, I was informed that a gentleman farmer there had raised a very fine plantation of Ash trees, on a piece of moor-land, which was worth little for any other purpose. Knowing him to be a very ingenious and capable farmer, both willing and able to communicate useful knowledge, I thought my examining the plantation, and giving you a just account of its planting, progress, produce, and present state, might be acceptable to the Society.

The soil was a black boggy moor, and had formerly been a hop ground; but so wet that it would not answer for that or any other purpose in agriculture, although it had been cut across with many open drains, five feet deep, to take off the water. The quantity was three statute acres, and the following account of the planting and produce was given me by the farmer in writing from his own register.

'In

'In the spring of 1764, I planted these three
'acres of black moor with small seedling Ash
'plants, drawn from my woods, hedges, and waste
'grounds, at four feet distance from each other.
'When they had stood two years, I cut them
'down within four inches of the ground. I then
'let them stand ten years, during which time
'they throve exceedingly; and in February 1776,
'I cut one acre and a half. The produce was as
'follows:

	£.	s.	d.
'31 hundred of poles, sold on the premises for	39	6	0
'11 loads of firewood, sold also on the premises at 16s. per load	8	16	0
	£.48	2	0

'The other acre and a half is still standing, and
'much superior to that already cut.'

On examining the standing part of the plantation it appeared to be in the most healthy and vigorous state. The shoots were generally three in each root, strait and clean; the bark being clear, smooth, and of a fine blueish green. The annual shoots were frequently from three to four feet in length; and from their present appearance I am fully of the opinion that, if cut next spring, this

part

part of the plantation will exceed the other at least one-third in value. The young shoots, in that part of the plantation cut in 1776, are remarkably strong and healthy, and bid fair to be fit for a second cutting in seven years.

By this easy and judicious management, one acre and a half of land, not worth five shillings a year for any other purpose, has paid the planter near fifty pounds in twelve years; and the acre and a half now standing will, if cut next season, probably bring him full seventy pounds.

The first five years after planting, they were kept clear from weeds, but that trouble and expence has been long since at an end; and in time to come, after deducting the small charge of cutting, the whole produce may be reckoned clear gain.

This, surely, must be a sufficient encouragement for gentlemen in other counties to plant such lands in the same manner, as it will at once prove beneficial to the owner, and to the community.

I am, &c.

EDMUND RACK.

Bath, June 28, 1779.

Article XLIV.

On the Use of Stagnant Water as a Manure.

[By a Gentleman Farmer in Norfolk.]

Gentlemen,

I Make no apology for transmitting to you an account of the following experiment, because I think it may prove as useful to others as it has been to myself.

At the lower end of my farm-yard is an old pond, or reservoir of water, which is the common receptacle of every thing that drains from the other parts of my yard, stables, and the ditches of several fields. On my first coming into the farm, it was nearly dry, but during the course of the winter, a considerable quantity of water was collected in it, which, as the spring advanced, grew very thick and dark-coloured, and in the summer abounded with insects.

The weather proving dry, and my pumps failing, I used a quantity of this water for my garden, and was soon surprised to see how strong and vigorous the plants proved that were watered with it. This led me to consider that it might probably prove a rich manure for pasture-land; but the quantity then

remaining

remaining in the pond was too small to make any considerable trial with. Determined, however, to ascertain the truth or fallacy of my conjecture, in the latter end of July 1772, I measured out two spots of fifty square yards each, in an adjoining meadow, which had been mowed, and was much burnt up. And in order to prove how far this exceeded other water, I watered one spot with it, and the other with water from a small adjacent river, three times a week, for a month together, there being little rain all that time. I observed the effects carefully, and at the end of the month, the two spots were in the following state:—That which had been watered from the river was far better than the rest of the field. The grass was tolerably thick and high, but weak and faint, seeming to have little virtue in it, and of a yellowish green. But on the other spot, which had been watered with the pond-water, the grass was much thicker and higher; being as strong and succulent as any part of the first crop had been, of a deep healthy green, and near eighteen inches high. I then determined to cut both, and keep them separate, in order to ascertain the comparative value of the hay. I did so; and when it was made, on weighing each, I found that on which the pond-water had been used near double in quantity, and much superior in quality, to the other.

I did not water either of the spots any more that season; but the next summer, I found the effects of this watering to an inch in the said meadow; the grass being much thicker and higher than on any other part of the field. I considered this as proof positive, and determined in future to increase the quantity of my pond-water. For this purpose I emptied the pond, enlarged it, and lined the bottom and sides with clay eight inches thick, to prevent the water from soaking into the earth. I then laid covered drains into it from my stables, ox-stalls, kitchen, dairy, and necessary, (the latter I regularly emptied once a year into it) and threw in all the offal made in the house, cabbage-leaves, rotten fruit, and the like; by these means the water soon grew very putrid, and I had it in great plenty. In my garden I now used no other manure, and yet found the produce much superior to my neighbours, who dunged ground equally good freely. Having a common water-cart made with a trough behind full of holes, I then watered my pasture and meadow land with the greatest success. Twenty carts of this water on an acre in the beginning of May, and in July, would render my crops of grass and rowen* far better than any manure I could lay on without it. After this success, I tried

⁕ Aftermath.

it on arable, and found it equally ferviceable on corn as on grafs lands.

If this relation of my experience fhould induce any of the Weftern farmers to follow my example, they will have no caufe to repent their labour; and it will fufficiently recompence me for the trifling tafk I have undertaken in communicating this to you.

I am, &c.
R. S.

W——, *July* 10, 1779.

Article XLV.

On the Management of Clover.

[By a Gentleman Farmer in Suffolk.]

Gentlemen,

AS clover is a grafs which fuits our climate better than almoft any other, I think the proper cultivation of it an object of national importance; and therefore take the liberty of informing you, how I have for feveral years managed it with great fuccefs.

In April, after my barley is come up, I fow about eight pounds of clover-feed per acre on it, and roll the land. This anſwers two good purpoſes, namely, preſſing and covering the ſeed, and fixing the roots of the barley more firmly, which, in a light ſoil eſpecially, is of great ſervice.

After the corn is reaped, I omit turning in any cattle till the crop of clover gets up pretty high and thick, which it will generally be by the end of October. I then turn in ſheep and other ſmall cattle for about a month, or, if the crop be large, ſix weeks. After this time, I let it remain unfed till April. My cattle are then turned in, which eat it off pretty bare by May, at which time I clear it for a crop of hay. If the ſeaſon is not remarkably dry, it will be ready for the firſt cutting by the middle of June, and generally yields me two tons per acre.

Experience has taught me, that the nearer the ground clover is cut the better, if it be cut early; but if it has ſtood too long, the bottoms of the ſtalks will be dry and naked.* In that caſe, it ought not to be cut ſo low, as the hay would be more ſticky and coarſe. If rain follows the firſt

* To this the writer might have added, that the ſtalk being drained of its moiſture, the root is alſo much exhauſted, and will require longer time before it ſends forth new ſhoots for a ſecond crop.

cutting,

cutting, the second crop will be ready about the 10th of September. The best time is when the flowers are all full blown, and the earliest begin to turn brown.

When I intend the second crop for seed, I usually let it stand till near October. This occasions it to thresh the better, and there is no danger of the seed shedding in the field.

In order to prevent the inconveniences that seed clover is liable to in a wet autumn, I generally leave half my crop unfed in April, by which means it is fit for cutting near a month sooner than it otherwise would be; and the second, or seed-crop, is brought more into the summer. When the autumn proves wet, this method is attended with many advantages; the seed ripens better, and is threshed with much less trouble and expence.—The sample is also better coloured, and the straw, being less beaten to pieces, makes better fodder for my cattle.

The best method I have ever found to prevent cattle from being hoved, as it is here called, or choaked, on their being turned into green clover, is to let them remain at the first no longer than till their bellies are full; and while feeding, to keep them

them constantly stirring. For as it is their greediness in swallowing the mouth-fulls too fast, and before it is sufficiently chewed, which occasions these accidents; if they are interrupted every two or three mouth-fulls, so as to give time for the balls to sink into their maw before the next follows, it will effectually prevent suffocation.——Whenever, notwithstanding this precaution, any of my cattle have swelled, I have immediately opened a vein, and stabbed them in the flank near enough the hipbone to prevent wounding the intrails. As soon as this was done, I put a quill or reed into the orifice to keep it open, that the wind might have a free passage out, and keep the animal warm till it recovered its breath. By this means I never lost more than one, and that was occasioned by the remedy being applied too late.

I am, &c.
W. E.

Near Halfworth, May 20, 1779.

ARTICLE

ARTICLE XLVI.

Thoughts on the ROT *in* SHEEP.

To the Secretary of the Bath Agriculture Society.

SIR,

THE great attention of the Bath Society to such subjects as promote the publick good, induces me to trouble you with a few loose thoughts relative to a disorder most fatal in an animal of vast importance to the " agriculture, manufactures, " and commerce of this kingdom."

The cause of the Rot in Sheep, says Mr. BOS-WELL, in his late ingenious Treatise on watering Meadows, is unknown.—Mr. ARTHUR YOUNG, in recapitulating all the information he could get in his Eastern Tour, observes, that " the accounts are so amazingly contradictory, that nothing can be gathered from them," but concludes, that " *every one knows* that moisture is the cause."

In differing from an author of Mr. YOUNG's acknowledged merit, supported by the general opinion of mankind, I am led to examine my own sentiments with caution and distrust;—but unless it is only meant, that moisture is *generally the remote cause,*

cause, it will be difficult to account for the Rot being taken on fallows in a single day, and in water-meadows sometimes in half an hour, when in grounds of a different sort, although excessively wet and flabby, sheep will remain for many weeks together uninjured.

Another opinion, which has many adherents, is, that the rot is owing to the quick growth of grass or herbs that grow in wet places.

Without premising that all-bounteous Providence has given to every animal its peculiar taste, by which it distinguishes the food proper for its preservation and support, (if not vitiated by fortuitous circumstances) it seems very difficult to discover, on philosophical principles, why the quick growth of grass should render it noxious,—or why any herb should at one season produce fatal effects, by the admission of pure water only into its component parts, which at other times is perfectly innocent, although brought to its utmost strength and maturity by the genial influence of the sun. So far from agreeing with those who attribute the Rot to quick-growing grass, which they call flashy, insipid, and destitute of salts, to me the quickness of growth is a proof of its being endued with the most active principles of vegetation, and is *one* of

the

the criterions of its superior excellence. Besides, the constant practice of most farmers in the kingdom, who with the greatest security feed their meadows in the spring, when the grass shoots quick and is full of juices, militates directly against this opinion.

Let us now consider whether another cause may not be assigned, more reconcileable with the various accounts we receive of this disorder. If our arguments, however specious, are contradictory to known facts, instead of conducting us in the plain paths of truth, they leave us in the mazes of error and uncertainty.

Each species of vegetables and animals has its peculiar soil, situation, and food, assigned to it. Taught by unerring instinct " the sparrow findeth " her a house, the swallow a nest, and the stork in " the heavens knoweth her appointed time."— The whole feathered tribe, indeed, display a wonderful sagacity and variety in the choice and structure of their habitations. Nor can it be doubted that the minutest reptile has its fixed laws, appointed by Him whose " tender mercies are over " all his works."

The

The numerous inhabitants of the air, earth, and waters, are strongly influenced by the seasons, and by the state of the atmosphere; and the same causes, perhaps, that rapidly call myriads of one species into being, may frequently prove the destruction of another. Is it then improbable, that some insect finds its food, and lays its eggs, on the tender succulent grass found on particular soils, (especially wet ones) which it most delights in? Or that this insect should, after a redundancy of moisture, by an instinctive impulse, quit its dank and dreary habitation, and its fecundity be greatly increased by such seasons, in conjunction with the prolific warmth of the sun?

The flesh-fly lays her eggs upon her food, which also serves to support her future offspring; and the common earth-worm propagates its species above ground, when the weather is mild and moist, or the earth dewy.

The eggs deposited on the tender germ are conveyed with the food into the stomach and intestines of the animals, whence they are received into the lacteal vessels, carried off in the chyle, and pass into the blood; nor do they meet with any obstruction until they arrive at the capillary vessels of the liver. Here, as the blood filtrates through the extreme branches,

branches, answering to those of the Vena Porta in the human body, the secerning vessels are too minute to admit the impregnated ova, which, adhering to the membrane, produce those animalculæ that feed upon the liver and destroy the sheep. They much resemble the flat fish called plaice, are sometimes as large as a silver two-pence, and are found both in the liver, and in the pipe (answering to that of the vena cava) which conveys the blood from the liver to the heart.

If the form of this animal be unlike any thing we meet with among the insect tribe, we should consider that it may be so small in its natural state as to escape our observation. Or might not its form have changed with the situation? " The ca-
" terpillar undergoes several changes before it pro-
" duces a butterfly."

The various accounts which every diligent enquirer must have met with (as well as the indefatigable Mr. YOUNG) seem very consistent with the theory of this disorder.

If dry limed land in Derbyshire will rot in common with water-meadows, and stagnant marshes; if some springy lands rot when others are perfectly safe;—is it owing to the circumstance of water, or or that of producing the proper food or nidus of the insect? Those who find their after-grass rot till the autumnal watering, and safe afterwards, might probably be of opinion, that the embryo laid there in the summer is then washed away or destroyed.

With regard to those lands that are accounted never safe, if there is not something peculiar in the soil or situation, which allures or forces the insect to quit its abode at unusual seasons, it may be well worth enquiring, whether, from the coarseness of their nature, or for want of being sufficiently fed, there is not some grass in these lands always left of a sufficient length to secure the eggs of the insect above the reach of the water?

Such who assert that *flowing* water alone is the cause of the Rot, can have but little acquaintance with the Somersetshire clays, and are diametrically opposite to those who find their worst land for rotting cured by watering. Yet, may not the water which produces this effect be impregnated with particles destructive to the insect, or to the tender germ which serves for its food or nidus?

For solving another difficulty, " that no ewe " ever rots while she has a lamb by her side," the
gentlemen

gentlemen of the faculty can best inform us, whether it is not probable, that the impregnated ovum passes into the milk, and never arrives at the liver. The same learned gentlemen may think the following question also not unworthy their consideration:

Why is the Rot fatal to sheep, hares, and rabbits, (and sometimes to calves,) when cattle of greater bulk, which probably take the same food, escape uninjured?

Is the digestive matter in the stomach of *these* different from that of the others, and such as will turn the ova into a state of corruption; or rather, are not the secretory ducts in the liver large enough to let them pass through, and be carried on in the usual current of the blood?

It seems to be an acknowledged fact, that salt-marshes never rot. Salt is pernicious to most insects. They never infest gardens where sea-weed is laid.* Common salt and water is a powerful expellent of worms bred in the human body.

* And yet sea-weeds, steeped a few days in the purest spring water, abound with animalculæ of various species.

I could wish the intelligent farmer would consider these truths with attention, and not neglect a remedy which is cheap and always at hand.

Lisle, in his book of husbandry, informs us of a farmer who cured his whole flock of the Rot, by giving each sheep a handful of Spanish salt, for five or six mornings successively. The hint was probably taken from the Spaniards, who frequently give their sheep salt to keep them healthy.

On some farms, perhaps the utmost caution cannot always prevent the disorder. In wet and warm seasons, the prudent farmer will remove his sheep from the lands liable to rot. Those who have it not in their power to do this, I would advise to give each sheep a spoonful of common salt with the same quantity of flour, in a quarter of a pint of water, once or twice a week.

When the rot is recently taken, the same remedy, given four or five mornings successively, will in all probability effect a cure. The addition of the flour and water will, in the opinion of the writer of this, not only abate the pungency of the salt, but dispose it to mix with the chyle in a more friendly and efficacious manner.

Were

Were it in my power to communicate to the Society the result of actual experiment, it would doubtless be more satisfactory. They will, however, I am persuaded, accept of these hints, at least as an earnest of my desire to be serviceable. Should they only tend to awaken the attention of the industrious husbandman, or to excite the curiosity of some other enquirer, who has more leisure and greater abilities, I shall have the satisfaction of thinking that my speculations, however imperfect, are not entirely useless.

I am, Sir,

Your very humble servant,

BENJAMIN PRICE.

Salisbury, Dec. 3, 1779.

ARTICLE XLVII.

On the Mode of Cultivating and Curing the Rheum Palmatum, or True Rhubarb.

GENTLEMEN,

AS the true Rhubarb of the shops is a very valuable drug in medicine, and, considered as an article of foreign commerce, very expensive; I would beg leave to propose to the Society at Bath,

Bath, that they endeavour to encourage its cultivation in this country, by offering a premium to the perſon who ſhall raiſe the greateſt quantity of the beſt kind, and cure it in ſuch a manner as to render it equal in quality to that annually imported from abroad. It is a plant to which our climate is not unfriendly, and it may eaſily be cultivated with ſucceſs.

I had laſt ſummer ſome plants of it in my garden, which were very vigorous, riſing to the height of eight feet. The roots weighed from eight to twelve pounds, and, when cured, the quality was allowed by the faculty to be equal to that of the Turkey Rhubarb.

But to give an opportunity for the medical gentlemen of Bath to examine and aſcertain its quality, I herewith ſend you a ſpecimen of the cured root, and a quantity of the ſeed in good preſervation, for ſuch perſons as may chuſe to cultivate it.

As to the culture of this plant, my knowledge has been chiefly acquired by my own experience; and that it may become an uſeful article of agriculture and commerce is, I think, not a very hazardous preſumption.

The

The feeds fown upon a very gentle hot-bed in March, readily vegetate; and when the roots are about the fize of a crow's quill, they fhould be carefully drawn up to preferve the tap-root, and planted in fine rich earth in a deep foil; and if the weather fhould prove dry, they muft be watered. When the plants are once in a growing ftate, all further care and trouble, but that of keeping them free from weeds, is at an end.

The diftance of the plants from each other fhould be eight feet; and as they difappear about feven months in the year, in this interval the ground may be ufefully employed in many articles of gardening, from the middle of Auguft to the beginning of April.

I am of opinion, the feeds will grow in the natural ground, if fown in a good expofure; but this I have not tried. It is, however, a little remarkable, that although innumerable feeds fall annually into the ground, I never perceived a fingle plant to grow fpontaneoufly.

The feed which I now prefent to the Society will afford opportunities for making various experiments, from whence fome certain principles may be drawn. The beft feafon for taking up the root for curing is, I think, when and as soon as the stem and leaves decay. If taken up in the spring, it is so succulent as to be dried with difficulty, and I believe loses a good deal of its resinous particles by the glutinous juice that issues from it.

The specimen sent herewith is from a root of six or seven years growth, taken up about a month since. It should, on taking up, be divided into proper parts, and the outer rind sliced off, then hung on a string exposed to sun and air, and defended from wet. Each piece should hang separate from the other, and care should be taken that it does not grow mouldy. When hardened on the outside, let it be removed to the corner of a kitchen chimney, where a moderate fire is constantly kept, till it is perfectly dry. It may then be rasped, and all the discoloured outside taken off.

I am convinced that the older the plant is, the better the quality will be;* for although it may have great virtues at four or five years growth, and may attain to upwards of twenty pounds weight when green, yet the root will be horny and flinty when dried, and not of that woody fine texture or appearance which it acquires at a more mature growth.

* See ARTICLE XLIX.

The

The ground on which mine was raised is a garden; the soil deep and fertile, but has not had any sort of manure since the seedling plants were first raised.

I would just add, that this plant does not seem fitted by nature for transplanting; and if it should be found capable of being raised in the natural ground, I think it would thrive much better.

If some such method as the following were tried, it might answer:—After marking out the ground at proper distances, take out the earth twelve inches deep and eighteen diameter; let the said earth be sifted and put in again loosely, then sow a few seeds thereon, and cover each plat with a hand-glass. If they succeed, the most central plants may be left to stand, and the rest drawn up.

    I am, &c.

            R. D.

Minehead.

            Article

ARTICLE XLVIII.

On the Cultivation of RHUBARB.

[By a Gentleman near Norwich.]

GENTLEMEN,

I AM greatly obliged to you for the favour of your letter, inclosing some seeds of the Rheum Palmatum, together with the directions sent by your ingenious correspondent at Minehead;* in return for which I have transcribed and sent you those which I sometime since received from a gentleman who has resided many years in Russia, and who assured me he received them from the late Dr. MOUNSEY, who was archiater to the Empress, and who had constantly followed the method here directed in that country. The Doctor's words are these:——

'The proper time for planting the seeds of the
'Chinese or the Turkey Rhubarb, is in April or
'May: they may be planted in flower-pots, three
'or four seeds in a pot, and plunged in a hot-bed
'until the seeds vegetate. When the plants are
'about two months old, let them be transplanted
'into the place where they are to remain, which
'should

* See preceding letter.

' should be in a fine light soil. It may not be
' improper to keep some of the plants in the pots
' till October, and some till the spring following,
' and then plant them out as above. When by
' these precautions you have secured a sufficiency
' of plants, you may afterwards venture to sow
' your seeds in the open air, as I have constantly
' done with success. If the seeds vegetate late in
' the season, they ought to be covered with mulch
' or moss, to preserve them in winter. When
' transplanted, set them at least four* feet asunder
' in the quincunx order, or in square rows; hoe
' them and keep them clean from weeds, and let
' the ground between each row be turned up yearly,
' taking care not to touch the roots. In the second
' or third year, the plants will begin to bear seeds,
' which you may sow at various times after their
' maturity, till you find which season suits them
' best. The earliest period at which the roots are
' useful, is at four years' growth, but even then
' they will be soft and spungy. So that unless for
' curiosity, or through necessity, they should remain

* This, in our climate, is not a sufficient distance—they should be eight feet apart; and even then, if the plants are strong and vigorous, the leaves will meet each other. This we assert, from our own knowledge; and are of opinion, that in a rich soil, if they were left ten feet apart, the roots would be still larger and better.

' eight

'eight years undisturbed,* although still more years
' will add greatly to their perfection. The roots
' are to be taken up in autumn after the stems and
' leaves are withered and decayed, but the planter
' may take them up in every season of the year,
' when he has a sufficient number, as it is uncertain
' at which season the roots will prove most solid.
' Upon taking them up, split them into two or three
' pieces, and hang them upon cords or rods in a
' kitchen or room with a stove in it, that they may
' dry with a gentle heat.'

Thus far Dr. MOUNSEY.

Some few gentlemen in Norfolk have, for their amusement, cultivated Rhubarb; they have planted the Rhaponticum, the Palmatum, and the Compactum, and managed their plants very much as above-directed. A near relation of mine, who is a physician, has used the Rhubarb of his own growth some time, and pronounces it as good as any foreign Rhubarb. He had some plants of the

---

* We apprehend that this part of the Doctor's direction must have been occasioned by the difference in climate between England and Russia. The latter is not so favourable to vegetation as the former. Perhaps four years here may bring this plant to as high a state of maturity as eight in Russia; and we are the more induced to think this is the case, from having seen plants raised in this country, the roots of which when properly cured, at four years' growth, were equal in quality to Turkey Rhubarb.

Palmatum

Palmatum and Compactum standing so near together, four or five years ago, that the seeds saved from them produced only *mule* plants, the roots of which, he believes, will prove as good and efficacious as those of the best original plants. They are now growing in his garden within a mile of Norwich.

The following is his account of the discovery and introduction of the different sorts into medicine; which I flatter myself may not be unacceptable:
'The Rhapontic was the *rha* or *rheum* of Dioscorides, and all the ancient Greeks and Romans. The Turkey and India kinds were utterly unknown to them. The Rhapontic was long supposed to be the true Rhubarb, till the discovery of the *Undulatum* about eighty years since; which was looked upon to be the true officinal Rhubarb for half the present century. It was then discarded for two competitors, the *Palmatum* and the *Compactum*, to both of which the preference has been given by different persons. Linnæus asserts the Palmatum to be the true Turkey Rhubarb; Mr. Miller the Compactum.'

I have now about twenty plants of the Rheum Palmatum of one year's growth only. They were raised in a box of good mould, set upon one of the borders

borders in my garden, and planted out at two months old where they are to remain. They appear to be strong and vigorous. I have also some from the seed sent me by your Secretary; and others from a friend in Russia, sown in the open ground, where I have no doubt of their coming to perfection.

I am, &c.

Sept. 16, 1778.

ARTICLE XLIX.

On the Cultivation and Cure of the True RHUBARB.

GENTLEMEN,

PERUSING the Farmer's Magazine for September, I was much pleased with some judicious remarks on the cultivation of the Rheum Palmatum, made, I think, by a gentleman at Minehead. With the utmost deference and respect to that gentleman's abilities and experience, I beg leave to submit the following observations, relating to the culture of that valuable plant, to your consideration; which may be depended on as facts, proved by myself and others from long experience.

The seeds of this plant do not require any hot-beds to make them vegetate; but if sown in the natural ground in the spring, when the weather is open, soon come up, and thrive very fast. It delights most in a rich, light, deep soil, and warm exposure; but will thrive in almost any soil or situation. If the roots be covered with litter, or the earth be drawn over them in winter, they will rise the stronger in the following spring. The seeds should be sown where the plants are to remain; and when they appear, the ground should be kept clean from weeds. When thinned out, the distance from plant to plant should be eight feet.

The above is all that is necessary to be understood by those who wish to cultivate this plant in perfection.

As to *curing* the root for medicinal uses, I must own myself a novice in the art, this being the first year I ever attempted it; and my roots being dry, I cannot with any precision say how they will turn out; but submit the following hints to your consideration:——

To have the root of a fine close grain or texture, drying it gradually seems to be essentially necessary.
I take

I take mine up, clean it from all dirt, and lay it in the shade, under a shed for two or three days, where, without becoming shrivelled, it will lose by degrees the exuberant moisture it had when recent from the earth. If it be exposed too suddenly to heat, either natural or artificial, or a very drying air, the root grows wrinkled, and is always horny or flinty.

Herein lies the chief difficulty; for when it is once well preserved thus far, it is safe: you may afterwards finish the process of curing, so it be done gradually, in any manner you may chuse, with success.

I cannot hold with barking or slicing the root, because, by the too easy admission of sun or air, great part of that resinous and glutinous matter, which I apprehend to be the richest part of the root, is drained off and evaporated; and which, under cover of the bark, would by degrees condense and harden with the root itself. You will, therefore, I am satisfied, find roots so cured to be of a much better texture, and richer quality, than those that are barked. The older the root, the better it is for curing.

I am, your's, &c.

Oct. 17, 1778.   G. P.

P. S. I forgot in its proper place to inform you, that that part of the root (for there are several buds, or **eyes**, which will bloom in future) from whence the main or any flower-stem issues, on perfecting its seed, immediately, or at least very soon after, begins to decay, and leaves the other buds sound, some or one of which will bloom the following season, according to their maturity. This is an undoubted fact; and therefore, although the older the root is the better it will be for curing, when it has passed its meridian, that proposition must be erroneous.

It is therefore my opinion, that the most proper time to take up the root for curing will be immediately on its perfecting the seed from its main or flower-stem, and to preserve *that part only*, every season, and to plant the buds with their respective roots again.

Article

Article L.

On the Cultivation and Cure of the True Rhubarb.

[Letter II. by G. P.]

Gentlemen,

YOUR Secretary's remarks on my objections to the flicing of the roots of Rhubarb when taken up to dry, feem very juft: but give me leave to obferve, that on my cutting the root, a quantity of matter, of the confiftence of melted glue, iffued from it, which, after the aqueous particles were extracted, hardened, and formed a gum or refin.

Query, Whether it is not neceffary to preferve as much as poffible of this fubftance with the root, as poffeffing a quality equal, if not fuperior, to the root itfelf? And allowing the root to be cut in lengths, but not barked or fliced, (which was my meaning*) would not a great deal of that refinous matter be thereby preferved; and would not the admiffion of air and fun, at the extremities of fuch pieces only, be fufficient to extract the aqueous particles, and to purify and concoct the juices?

* See page 194.

As to his second remark on the decay of that part of the root from which the flower-stems arise, he might possibly have never made the observation. This is the first time of my observing it, though I have at divers times heard it attested by those who have experienced it. In the specimen I send you, the side-bulbs are apparent, and the main root in a state of decay.

It seems to me that this plant, like many other perennials, upon its first bloom, exhausts (if the bloom be strong) its vegetative principle in that part. I am sensible the root cannot be cured without shrinking considerably, and it always appears much-shrunk when cured with the bark on. If it be exposed to heat, or a drying air, when first taken up, it will shrivel very much, and be horny; to prevent which, I advise laying it in the shade, or under cover.

I am, &c.

G. P.

Article LI.

Reply to some Enquiries relative to the True Rhubarb.

[By Dr. John Coakley Lettsom, F. R. A. S. S.]

Gentlemen,

THERE is every reason to conclude with Linnæus, that the Rheum Palmatum is the Turkey or Russia Rhubarb.

The root is perennial, but throws out annually, from its crown and sides, new shoots or bulbs, which flower and decay in succession.

It may probably be of little consequence, as to the vigour of the roots, whether they are taken up in summer or autumn; but as warm weather is best for drying them, the former seems most eligible.

The roots, if large, should be sliced, so as to admit of a free exsiccation.

I believe Rhubarb delights in a sandy soil, on a somewhat elevated situation; such a soil as carrots will flourish in.

Dr.

- Dr. HOPE, of Edinburgh, has paid great attention to this exotic, and is very capable of giving its history, a very essential part of which [*i.e.* its medicinal powers] is yet uneſtabliſhed;—to him, therefore, I refer you.

And am, &c.

J. C. LETTSOM.

London, Nov. 21, 1778.

ARTICLE LII.

Anſwers to Queries from the Bath Society reſpecting Rhubarb.

[By Dr. HOPE, of Edinburgh.]

GENTLEMEN,

1. THE Rheum Palmatum is the Turkey, or Ruſſia Rhubarb. The India Rhubarb is the root of another ſpecies, or variety.

2. I believe your obſervation reſpecting the root dying at four years old to be well founded; and therefore it ſhould be raiſed at that age.

3. I am of opinion, that the entire root ſhould be hung up in the open air for two, three, or more weeks,

weeks, and thereafter cut into marketable pieces before it be put into the drying.——N. B. The cuticle should be rubbed off as soon as possible.

4. The succulent root is more purgative than the dried, therefore the more recent the better.

5. A soil that is fit for carrots will suit Rhubarb. It is believed that the roots raised in a dry soil are preferable to those raised in a moist one.

6. The season for taking up Rhubarb is from the end of July to the first of January; it should be taken up after the weather has been some time dry.

7. I know nothing to prevent its being transplanted.

I have the honour to be,

your most obedient servant,

JOHN HOPE.

ARTICLE

Article LIII.

On the Growth and Application of Rhubarb.

[By a Gentleman near Norwich.]

Gentlemen,

I Herewith inclose you an account of the growth and application of some Rhubarb, sent to me by a particular friend, who is a physician of very extensive practice, and on whose accuracy and integrity I can fully rely; in which you will perceive his experiments, as to the weight and size of the roots, entirely militate with the idea one of your correspondents adopted; 'that in consequence of ' the annual decay of that part of the root which ' corresponds with the flowering stem, the roots are ' in fact never more than four years old, that being ' the usual time of their flowering.'

For this difference of opinion, if I may hazard a conjecture, I should endeavour to account, by suggesting, that although that part of the bulb or root from which the flower-stem arises may decay, yet that the fangs or tap-roots of the plant do not decay with it, but increase annually for several years. My friend the physician has, you will observe, an idea of its becoming useful as a dye, which is, I believe,

I believe, new, and may, if properly pursued, prove of much importance; but take his account in his own words:—

" In the summer of 1771, I had a plant of the true Rheum Palmatum, in great vigour and in full flower, growing at the distance of about four yards from a plant of the Rheum Compactum, which was likewise at the same time in flower. As the first was the plant generally allowed to be the true Turkey Rhubarb, I carefully collected and preserved the seeds of it, which I sowed early in the spring of the year 1772, in a bed of common light earth, about half an inch deep. In about five weeks the plants appeared in great plenty, and were, in the beginning of the winter following, thinned, and transplanted at the distances of six and seven feet from each other. The plants were healthy and strong, although they had no particular care or attention paid to them. But what seemed most remarkable in them was, that the leaves were neither those of the Palmatum or Compactum, but a perfect mixture of both; very large and broad like the Compactum, but terminating in long sharp points, and in some degree indented, and resembling the Palmatum. In the summer 1775, they were all in flower, the stems being six and seven feet high:

high; when their seeds were ripened, they were carefully gathered, as they have been every year since, and regularly and constantly sowed every spring, but without having ever produced a single plant. Many botanical gentlemen have viewed these plants, and all pronounced them to be mule plants, betwixt the Palmatum and Compactum species.

" In the winter 1776, I took up a root of these plants sown in 1772, and laid it in a south window to dry. It had several long perpendicular pyramidal roots, about nine inches in length, and better than half an inch in diameter. They shrunk very much in the winter, but were in the spring sufficiently dry to be reduced to powder. I gave different doses of it to divers persons with all the good effect of very mild Turkey Rhubarb, although the quantity was nearly doubled. In the beginning of the winter 1777, I took up another root; the increase of the size and quantity of roots was then very great. The weight of the root taken up in 1776 was only between eight and nine pounds; that of 1777 weighed full fourteen pounds: this root dried better, shrunk less, and in every respect, when dried and prepared, resembled more the true Turkey Rhubarb. The effects were likewise produced by much smaller doses; but it was not altogether

gether fo purgative as the Turkey Rhubarb of the fhops. I gave to two or three perfons this root in its frefh ftate, that is, *undried*; directing them to bruife about half an ounce, and boil it in half a pint of water, till reduced to one quarter of a pint. This had all the good and fimilar effects with the true fhop Rhubarb, infomuch, as I am greatly inclined to think, that the green root of the Rheum Palmatum may be ufed with fafety and effect. This root, when dried and prepared, yielded a good quantity of well-looking Rhubarb, which, when powdered, had all the appearance of fhop Rhubarb, but was milder and more grateful to the tafte. I took once a dofe of this Rhubarb myfelf, for a complaint in my ftomach; for which I had always before taken Turkey Rhubarb, and found exactly the fame relief I had ufually received, only with a lefs purgative effect in the bowels.

" In the beginning of this prefent winter 1778, I have taken up two more roots; the one weighed eighteen, the other twenty-one pounds, and I have little doubt of their virtue and efficacy having been improved, as well as their fize and quantity increafed; and I am of opinion, they will continue fo to do in all the above refpects for two or three years longer. Eight years old, I am inclined to think

from

from some observations, is the meridian of their perfection. In these mule plants, produced without doubt from the farina of the Palmatum and Compactum intermixing with each other in 1771, the root does not appear to have been at all affected either in reality or appearance, having accurately compared it with that of the genuine Rheum Palmatum. I tried these roots both fresh and dried with the shop Rhubarb, by an experiment of another kind. I infused them in a portion of water, and to the infusion, when strained, I added a few grains of salt of tartar, whereby I obtained a very beautiful red tincture, such as would be valuable for the purposes of dying a colour which at this time is so very expensive, and which, by this means, may probably be amply provided for by the use of this root when it is more generally cultivated, as with very little trouble and expence it may be. The soil in which my plants were raised is very light for about twelve inches deep, under which there lies a stratum of red sand of great depth."———Thus for the Doctor.

I am, Gentlemen, your's, &c.

Dec. 7, 1778.

ARTICLE

Article LIV.

On the Extirpation of Plants noxious to Cattle on Dairy and Grazing Farms; and the Cultivation of such as are wholesome and nutritive recommended; with some Hints on the breeding and rearing Milch Cows.

[By Mr. Benjamin Axford.]

Gentlemen,

THERE is no branch of agriculture which to me appears more important in itself, or to open a larger field for improvement, than the conducting and management of Dairy Farms. This will be very evident, when we consider it as a fact, that the health and good condition of milch cows, and all grazing cattle, depend in a great degree on the conduct and care of the farmer, in keeping his pasture-lands clear from weeds and plants of a noxious quality, and in stocking them with such as are healthful, salutary, and medicinal.

But the most essential and weighty considerations are, that the health and lives of mankind are, in some measure, dependant on the health and good condition of milch-cows; milk being a vegetable juice, partaking more or less of the good or bad qualities of the plants on which cows feed.

Milk,

Milk, and its produce, in cream, butter, cheese, and many of our luxuries, are conſtituent parts of our daily food, from the earlieſt to the laſt ſtage in life; conſequently, great care ought to be taken with reſpect to the food of animals, which furniſh us with ſo great and neceſſary a part of our ſuſtenance.

Granting the above premiſes, it is humbly conceived, that the attention of the Bath (and every other) Agricultural Society cannot be employed in any purſuit that tends more to the intereſt and health of mankind, than the increaſing the quantity, and improving the quality of cow's milk. The taſk may be arduous; but in proportion to the ſucceſs attending their endeavours will be the reward.

That cows are frequently diſeaſed, is a well-known fact: and, I believe, moſt gentlemen who keep cattle are convinced, that the diſeaſes generally proceed from unknown cauſes. Few dairies of cows remain a ſummer all healthy. Among many inſtances that might be adduced, I will mention one, which, in the ſummer 1777, came within my own obſervation. I was witneſs to the loſs of five cows, out of a dairy of only thirteen; and the moſt noted Cow-Leeches could not diſcover, or even gueſs at the diſeaſe or its cauſe.

I have

I have also reason to believe, that the milk of diseased cows is too often mixed with the rest, and made into butter, cheese, &c. If then it appear, that numbers of cows are diseased, and die annually, without the nature or cause of the disease being discovered, and of diseases to which this species of animals are not naturally subject; I presume it will be most reasonable to search for that cause in their food.

On inspecting pasture and meadow-lands in general, many noxious and poisonous plants will be found, and sometimes in considerable quantities. Of these kinds are, among others, the following: henbane, hemlock, the aconite, or deadly nightshade, and several species of dropwort; which, if taken in with their food by cows, &c. will generally cause disease, and sometimes death.

I am aware of the objection that may be made to this suggestion of danger to cattle from noxious plants, i. e. that *instinct* is a certain guide to almost every species of animals in the choice of their food. This is generally, but not unexceptionably, true. If cattle were at liberty to rove at large over extensive tracts of pasturage, with a plenty always before them to choose out of, there would be little danger; but when herds of them are confined

within

within narrow inclosures, where such noxious plants abound, and kept there till little that is green remains, I think it almost impossible but that some of the cattle must be disagreeably affected by such plants when they are eaten.

In proportion then as pasture-lands are cleared of these and other noxious plants, the danger is lessened; and a considerable advantage will be derived from such lands being, by this means, rendered capable of producing a larger quantity of wholesome herbage.

All neat beasts have a natural tendency to scouring and flatulent disorders. It is therefore a duty of the greatest importance to the farmer, to sow and plant in his pastures and hedges such herbs, in proper quantities, as are found to be the best remedies for these and such other complaints as cattle are most incident to. Among many that might be mentioned, the following herbs are very salutary; lovage, agrimony, carraway, and cummin.

The general produce of ant-hills in this country has often been (through mistake) supposed to be wild thyme; and as this herb is salutary in its nature, farmers have suffered these hills to remain in their

their pastures, from an apprehension that they furnished a medicinal repast to their sheep and cattle. But on a careful examination, any person may be convinced, that, in general, the produce of ant-hills is, a *little* of the wild thyme, (which I never observed to be touched by cattle) and a much larger quantity of poor small rushy four grafs, which is a very pernicious kind of food both for cows and sheep. The extirpation of ant-hills is, therefore, an essential part of good husbandry. If the contents of them are mixed with lime and mulch (long dung) to rot, they make an excellent compost for the same or any other land that wants manuring; and this method will prevent the ants from bringing it in heaps again, which they are known to do when the hills are only levelled by spreading the earth round them. The bare places left by the removal of ant-hills, are very proper for sowing the medicinal herbs and plants above-mentioned, or the small Dutch clover, marl-grafs, &c. which will spread in and greatly improve the land at a small expence.

To persons unacquainted with agriculture, or who have not considered the above matters, this essay may appear frivolous and uninteresting; but those who have observed the impropriety and bad husbandry

husbandry of suffering such plants to grow on their lands as should not be eaten; or if eaten will injure the farmer's interest, by hurting his cattle, will admire that such indolence and extreme inattention should so generally be found amongst us. To such, it will appear an extraordinary act of negligence, to suffer pasture-lands to remain overrun with hemlock, thistles, docks, rushes, &c. and our hedges and ditches filled with poisonous plants of several species. Equally surprising will the inconsiderateness of farmers appear, in suffering such quantities of nettles, thistles, &c. to remain on the sides of our publick roads, till their seeds ripen and are carried by the winds into all the adjacent fields, where they produce most plentiful crops the succeeding spring.*

Cattle, when confined in such foul inclosures till their food becomes scanty, will, (as before observed) through hunger, devour a considerable part of such noxious plants with the rest of the herbage.

* We cannot help expressing our entire concurrence with Mr. AXFORD's observations on this head. The evil he complains of is so great and so general, that it cannot be too much exposed or censured. In vain may the farmer clean his fallows, or hoe his crops; while this remains, all his labour will be lost—all his care ineffectual.

Obvious

Obvious as thefe inftances of bad management, and the many difadvantages attending them, are, the greateft difficulty feems to be, that of making farmers in general fo fenfible thereof as to induce them to purfue the above eafy plan for redreffing and removing them. I think it is out of the reach of premiums; but perhaps fome honorary reward might ftimulate the more intelligent to undertake fo neceffary a work; and I am of the opinion, that if a few would fet the example, others would foon follow, and in time it might become general.

Some gentlemen farmers are very curious in the breed of cows, and management of their ftock on dairy farms. Such will, doubtlefs, improve on any hints that may be communicated through the Bath Society.

Much depends on the choice of cows, and the care taken to mend their breed, and increafe their milk. Cows of a red and black colour are preferable to white, of which not more than one fhould be admitted in a dairy. Heifers, intended for breeding, fhould not be bulled till the fourth year. The third, fourth, and fifth calves are the moft robuft, and of courfe the beft to breed from.

A bull

A bull should be well fed, and kept from coition at least till the second, if not till the third year. His vigour lasts only two years.

In the choice of cows to breed from, see that they have eight or ten white teeth in their jaw, that the breast be broad, the tail long, the veins of the belly distinguishable, the brace of the navel large, a broad forehead, large black eyes, wide nostrils and ears.

The feeds esteemed the most salutary in promoting an increase of milk, are those of trefoil, sainfoin, angelica, pimpernel, cummin, and anise. About the walls of houses, and on the insides of hedges, sow lovage.

Since the foregoing remarks were written, a work, called MINUTES OF AGRICULTURE, has come under my notice, and serves to confirm my sentiments of the great benefit that would arise from clearing pasture-lands of noxious weeds, and storing them with such as are salutary and medicinal. The writer of this work says, that " on the 29th " of August a cow died of the red water, and that " on opening her, the maw was full of half-digested " vegetables, although she had not eaten for many " days." Again,

"August 17, 1775, An ox died suddenly in a field that had been eaten down. The farmer could not account for it."

"December 4, Two oxen and one cow died with scouring; one ox blowed; two bulls surfeited; and one cow had the red water: all died, and the writer cannot account for the diseases."

ARTICLE LV.

Account of the Culture of CARROTS; *and Thoughts on Burnbaiting on Mendip-hills.*

[In a Letter to the Secretary.]

SIR,

IN pursuance of the directions of the Society, I herewith transmit particulars of the culture, expences, and produce of my carrot crop, which you lately did me the favour of inspecting.

From a hearty wish to promote the publick-spirited designs of that most laudable institution, I have taken the liberty to annex a few remarks on the comparative advantages of liming and burn-baiting, in respect to the soil of the new inclosures on Mendip-Hills.

As the spirit of cultivating these wastes seems to overcome every obstacle, and as a knowledge of the inefficacy of burning may prevent much useless expence, I trust these few hints, drawn from real experience, will not be thought trifling or unimportant.

The field in which my carrots were raised was a few years ago part of the forest of Mendip. It contains eight acres: the soil a gravelly loam, of a good depth.

In the year 1776, it received an ample manuring with lime, (about twenty quarters per acre) and was sown with turnips; in 1777, with barley; in 1778, it was again manured with horse-dung, to the amount of fifteen cart-loads per acre, and planted with the large Scotch Cabbage. The produce of this crop was very great, being more than thirty tons per acre, and the stock I maintained with them would astonish the farmer unaccustomed to the cultivation of this plant. And here I cannot forbear recommending, in the warmest manner, the culture of this cabbage (in conjunction with turnips)

turnips) to every spirited Agriculturist, and particularly to those who keep large flocks of sheep. Every person in that branch of farming must have frequently experienced, during severe frost, and deep snow, great difficulty in getting at his turnips. Now this inconvenience would be entirely obviated by his possessing three or four acres of this plant; for their height and hardiness render them accessible and found at all times, and in the most severe seasons. I will not say that the produce will be equal in weight to a well-managed crop of turnips, but will be bold to affirm, that one hundred pounds of Scotch cabbage will go as far, in keeping or fattening horned cattle, as one hundred and fifty pounds weight of turnips.——But to return:

In the spring 1779, I began preparing for my carrot crop. Particulars as follows:

	£.	s.	d.
Feb. 15. First ploughing across the ridges of the cabbages, 4s. per acre —	1	12	0
March 1. First harrowing, 9d. per acre ——	0	6	0
April 15. Second ploughing, 4s. per acre —	1	12	0
20. Second (bush) harrowing, 9d. per acre — — — —	0	6	0
30lb. red Sandwich carrot seed, at 1s. per pound —— ——	1	10	0
	£.5	6	0

		£.	s.	d.
Brought over —		5	6	0
April 24. Sowing by hand in drills, one foot apart, and covering the seed, 13s. per acre		5	4	0
June 4. Hand-hoeing and thinning, 20s. —		8	0	0
October. Digging up, 30s. —		12	0	0
Carting home, cutting off tops, and securing		10	0	0
Rent of land		8	0	0
		48	10	0
The produce was 640 sacks, of 4 bushels each, valued at 3s. a sack —		96	0	0
Each sack weighed upwards of 200 pounds				
Nett profit of the crop	£.	47	10	0

Nearly 6l. per acre. Quantity of carrots, 8 tons per acre.

From experiments which I have made, I am fully persuaded that carrots are worth more than three shillings per sack, in fattening hogs.

I will now proceed to give proof of the inefficacy of burn-baiting when applied to the soil of Mendip-hills, drawn from real experience, and designed as a caution to those who may be disposed to adopt this mode of improvement.

Having frequently met, in different authors, with the most flattering and encouraging accounts of this-

this plan of cultivation, and also been an eye-witness of very large crops procured thereby, on *black moory* soils, I formed a resolution of trying the effects of the ashes thus procured in comparison with lime. For this purpose, I selected a field in which there was no apparent variation of soil. As it was a new inclosure, and had never been ploughed, the furze, fern, &c. with which it abounded, added to the turf, furnished more than three hundred bushels of very fine ashes.

I then divided the field, and spread the three hundred bushels on half of it, fully expecting the most beneficial effects from so ample a manuring. On the other half of the field I spread four hundred bushels of lime, and sowed the whole in one day with wheat.

On the coming up of the wheat, I was very attentive to the field, and not a little surprised to see the limed part assume the most lively and healthful verdure, whilst the other part appeared very weak and languid, insomuch that the difference was perceivable at a mile's distance.

The limed part maintained its superiority from that time to harvest; and on threshing, I found the produce of the limed part to be twenty-four bushels, bufhels, and the burnt part only fourteen bufhels per acre.

Befides, this was not the only difference, for in lefs than a month after harveft, the furface of that part of the field to which the afhes had been applied, was entirely covered with young furze, while the other part remained perfectly clean and free from it.

I have alfo tried burn-baiting as a preparation for potatoes, and have been equally difappointed and unfuccefsful.

Wifhing the fociety all the fuccefs which their generous attention and activity merit;

I remain, Sir,

Your humble fervant,

JOHN BILLINGSLEY.

Shepton-Mallet, Dec. 7, 1779.

'ARTICLE,

Article LVI.

Result of Experiments to ascertain the Advantage of cultivating Rhubarb.

AFTER the receipt of the several letters relating to the expediency and advantage of cultivating the *Rheum Palmatum*, or True Rhubarb, on a large scale in this country, the Society requested several medical gentlemen to make experiments on the specimens sent by their correspondents. These experiments were made with care and accuracy, and the result is contained in the following *Report*, which was sent to the Society by Dr. Falconer, of Bath. In consequence of this Report, premiums, to the amount of one hundred pounds, were offered for cultivating and properly curing this plant in the four counties.

Dr. Falconer's *Report.*

Rhubarb is the *Rhaved* of the Arabians; the *Rha Barbarum* of Alexander Trallianus; the *Rheum* of Paulus Agineta; the *Rheum Barbarum* of Myrepsus; the *Rha Barbarum Officinale* of Caspar Bauhin, and of the London Dispensatory. What it is of Linnæus, I cannot say. Dr. Lewis says, it is the *Rheum foliis subvillosis peliolis æqualibus,*
Linnæi,

Linnæi, Spect. plant.; and so says Mr. VOGEL. Now, this description is affixed by Linnæus to the *Undulatum*, which is not at present understood to be the true Rhubarb. On the other hand, Dr. RUTTY asserts the *Palmatum* of Linnæus to be the true Rhubarb; and I believe that opinion is now generally thought to be right by the best botanists and ablest physicians. It is called *Rha*, by the Tartars; and from thence is derived *Rha Barbarum*, as growing among barbarous nations. It is named Rha, from the river Volga, which is so called by the Tartars, near which it is cultivated. It was first mentioned by Alexander Trallianus, in the year 560, but appears to have been in use among the Arabs prior to that period.

In earlier times, the *Rhapontic* was thought to be the true Rhubarb, and spoken of as such by Dioscorides and Celsus; being the *Rheum* or *Rha* of the former, and the *Radix Pontica* of the latter.

The marks of its goodness are, to be perfectly dry and friable, yet with a good degree of hardness or solidity, and perfectly uniform in its substance. It generally comes to us in roundish pieces, with a hole through the middle of each, and is externally of a yellow colour, but that in foreign

Rhubarb

Rhubarb is often artificial. When cut, it is of a fine reddish yellow, variegated with lively reddish streaks, intermixed with white. When powdered, it appears of a bright yellow, and on being chewed, imparts to the spittle a deep saffron tinge. Its taste is rather acrid, bitterish, and somewhat astringent; its smell is lightly aromatic; when chewed, it seems gritty, as if sand were mixed with it.

The specimen of Rhubarb presented to the Society, and submitted to my examination, answered to all these qualities. I compared it with specimens of the best Turkey and East-Indian kinds. It was rather, though very little, less aromatic and resinous than the former; and had somewhat fewer of the reddish streaks through its substance, but was much clearer, and more distinctly marked, than the East-Indian.

In specific gravity, it was near the Turkey, and not so hard or heavy as the East-Indian. In taste, I could not distinguish it from the Turkey, except that I thought it somewhat, though very little, fainter. The tincture made with brandy was of a bright, clear, yellow colour, not distinguishable from the Turkey, but superior to the East-Indian. The infusion with water was also nearly, if not altogether, equal in colour, taste, and smell, to the

Turkey,

Turkey, and superior to the East-Indian.———The smell of the powder was not distinguishable from the Turkey, and superior also to the East-Indian.

I tried its purgative virtue in several instances: and another gentleman, to whom I gave some of it, tried it also in several other cases. We agreed perfectly in our account, that its operation was, in every respect, such as might be expected from the best foreign Rhubarb.

Finally; I think the specimens shewn to me are extremely good in their kind, very little (if at all) inferior to the best brought from Russia or Turkey, and fully sufficient to supply the place of foreign Rhubarb. W. F.

ARTICLE LVII.

Observations on the best Method of destroying Vermin, and preventing the destruction of young Turnips by the Fly.

I Beg leave to offer to the Society an account of a few trials I have made to prevent the destruction of seed, and springing grain, pulse, &c. by vermin of different kinds. I am not excited to write

write by motives of intereſt, or with a view to obtain honour; but wiſh to caſt in my mite for the promotion of uſeful knowledge.

I have for ſome years paſt left off trade, and taken a ſmall eſtate into my own hands, principally for my own amuſement and inſtruction in the operations of nature.

As I am fond of a garden, I have frequently attempted to raiſe early peaſe; but was often diſappointed by their being deſtroyed by mice. To remedy this, I conſidered that *ſweets* were their delight, and judged that bitters might be their averſion; accordingly I ordered the gardener to ſteep both my peaſe and beans in common water three hours; and after ſowing them in drills as uſual, to ſhake ſome coal chimney ſoot over them pretty thick before he covered them in; by which means I have not loſt any for ſeveral years; and the ſoot proves a good manure.

Soon after I took the farm, I found I had many enemies to encounter with; ſuch as the *black flea*,* *grub-worms*, *birds*, *rats*, &c. I generally ſow cabbage-ſeed enough to plant out two acres of land,

---

* By the ſubſequent part of this letter, it appears, that by the *black Flea*, our author means the *Fly* that preys on young turnips.

for

for the support of my ewes and lambs* in the spring, when grass is scarce. But when I first took my farm, after several sowings, I could scarcely raise enough for my purpose, the black flea eating them off while young; but considering that this insect loves to be in the sunshine, I sowed my seed under the shade of apple-trees, and was not disappointed. This last summer twelve-month I attacked them in the sunshine, by sowing the seed in the garden, and scattering soot on the ground directly, by which means all my seedling plants were saved. This last summer, absence from home prevented my repeating the experiment. It is, however, an easy trial for gentlemen to make, and I think it may be depended on as an effectual remedy.

Chaffinches are also very eager in preying on seedlings soon after they are out of the ground, pulling them up by the roots, although they only eat the seed leaves. But whether they would attack them on ground that has been sown with soot, I cannot from my own experience say—the experiment is, however, easy and worth making. I think the best mode of trying it would be, while the dew is on the ground, for some person to

* This Gentleman seems not aware how good and profitable cabbages are early in spring for oxen.

powedr

powder over the seedling plants lightly with soot, before the third leaf appears. It is probable that the bitter flavour of soot is very difguftful to birds as well as infects.

The firſt year I ſucceeded very well in planting out my cabbages; the weight of the crop being from ten to twenty pounds, which was equal to my expectation, confidering the ground was light and fandy.

The next year, I fowed my cabbage-feed as ufual. When the plants were fit for tranfplanting from the feed bed, I attended, and found many of them had knobs or warts on the roots, fome the fize of a pea, and others larger. On cutting fome of thefe knobs off, I found a very fmall worm inclofed. I ordered them to be planted out; and looking over the ground after they had formed pretty good heads, I obferved many of them looked fickly, having a blueifh caft on the leaves. I pulled feveral of them out of the ground, and found the roots fwelled as big as a child's fift, the grub-worms being then come to perfection.

To prevent this, tranfplant only fuch as are quite clean from warts. At the time of planting, the ground is frequently dry, and fometimes even

ſcorched

scorched with heat. In that case, let the planter, after making a hole with his dibble, pour in some water, and stir it in till he has made the earth a soft puddle.* A boy should dip the roots of the plants first into water, and then into dry soot immediately before they are planted. According to my idea, this will prevent the grub from ever touching them.

I should now follow the grub-worms and birds into corn fields. The two first years of my sowing wheat, I could not get, from nine to ten acres produce, more than ten bushels per acre. On a nice examination, I found the grub-worms attacked my wheat *under* ground, and birds of various sorts *above* it. It was necessary to seek for a remedy. I ordered two bushels of seed wheat to be put on the barn floor as usual, with a proper quantity of lime and sea-water: [some use brine.] I then ordered a quarter of a peck of soot to be added, and worked well in with the corn, that all might be rendered bitter by it. If a larger quantity of soot be used, the better, as it proves good manure. My success the first year was, that I had twenty bushels on an average per acre. This last summer

* This is certainly a good method, and may be practised in a *Garden*—but when a Farmer wants to plant several acres, the process would be too tedious, if not quite impracticable.

the

the produce was still larger. After sowing, my servant told me, that if I did not send a person to frighten the birds, the wheat would be half carried away; however, I let it alone, to see the effect of the soot. Pheasants and partridges had scraped the ground very much. I particularly marked the spots, and at harvest, found the corn thicker there than in other places. This convinces me, that the birds could not find any corn suited to their palates, the soot having rendered it very bitter; and I had a good crop for so light a soil.

2*dly*. The crops of young turnips are frequently destroyed by the black flea [fly] notwithstanding many things have been tried to preserve them; all of which I am informed have failed.

On this subject I will offer a few thoughts:— The sense of smelling in the black flea [fly] and in all other insects, is exquisitely acute; without it they know not one plant from another, as their sphere of vision is probably confined to a very few inches. It is by this sense that they are guided to their proper food. The only thing necessary then is, to overpower the sweet smell they are attracted with, by one that is strong, fœtid, and disagreeable.

What

What I would therefore propofe is, that an acre of turnips be fown in the ufual way, and after the ground is finifhed, for the feedfman to throw on a peck or more of dry foot, as regularly as he fows the feed. If I am not very much miftaken, this will banifh or deftroy all the black fleas, [flies] and by that means fave the crops.

3*dly*. When firft I came to the houfe I now inhabit, we were much troubled in the night by the noife of rats. Mentioning this circumftance to the farmer, who was about to leave the premifes, he told me they had done great damage in the barns and corn-ricks. In the fide of a bank which ran along the path-way to my barn, I obferved a number of holes, in which the rats harboured and bred in warm weather. The next day making fome matches with flips of brown paper dipt in brimftone, I put them into the holes—the mouths whereof I ftopt, to drive the fmoak inwards. After the matches were burnt out, my man opened the ground, where we found feveral nefts; but both old and young were fled. The rats left my houfe, barn, and ftables, directly; and for five years paft have never returned. The difagreeable fmell of the burnt brimftone, doubtlefs, occafioned their leaving the premifes. I would therefore propofe, that when a barn is cleared out juſt before harveſt, a pan of charcoal be lighted up in it, and ſome pieces of brimſtone thrown on the fire, to fumigate the whole barn. If the doors and windows are ſhut cloſe, this will be done moſt effectually.

4*thly.* Having a field overrun with moles, I had the freſh mole-hills ſhovelled off, and the holes opened; I burnt a match in each, keeping in the ſmoak. The conſequence was, that all the moles left the field directly, and returned no more. But ſhould they return in future, it is only repeating the experiment, and I doubt not but it will have the ſame effect.

I ſhall now conclude for the preſent; but may probably employ ſome future hours in ſending ſuch other obſervations as may occur in the courſe of my experience; and which may be uſeful in promoting the laudable purpoſes intended by your inſtitution, to which I wiſh all poſſible ſucceſs.

And am, with great reſpect,

yours, &c.

J. JACOB.

Arne, near Wareham, Dorſetſhire,
 Jan. 20, 1780.

ARTICLE

Article LVIII.

On the Culture of Carrots, and the Rot in Sheep; by a Gentleman near Norwich.

[In a Letter to the Secretary.]

SIR,

I Thank you greatly for the two letters you transmitted to me in your last; the one from Mr. BILLINGSLEY, on the culture, expences, and produce of a crop of Carrots; the other containing Mr. PRYCE's thought on the Rot in Sheep.

On the first subject, as somewhat informed by having sometimes grown four or five acres, never less than one or two in every year, for a considerable time, I dare venture to assure you, that Mr. B's statement of the expences incurred is, in the articles of ploughing, harrowing, carting home, and securing, rather beyond the price I have ever paid for the same; and that he is not in any other of his articles beneath the fullest price here, nor is his produce greater than common, or what may be expected from such management; so that it is both as just and accurate an account as I have met with. The only objection to it which strikes me is, the heavy

heavy expence he was at of thirteen shillings per acre for sowing and covering the seed in drills, which practice, from two trials only which I have made of it, appears to me not so eligible as sowing the seed broadcast. The two drilled crops were with me the worst I ever grew. The seed of carrots, although ever so well rubbed with sand or any other substance, will still adhere together to that degree, as to render the delivery of it in drills not only tedious, but very uncertain; and wherever it falls in patches, the loss of ground is considerable; besides, the difference of nine-pence or a shilling per acre at most for random sowing, and thirteen shillings per acre for drilling, is an object worth attending to.

The great expence attending the culture of carrots, being the chief, perhaps the sole obstacle to the general growth of them, every abridgment of that expence should be studied, as it bids fair to promote their general use.

Perhaps the method in which I have for some few years past proceeded, where it can be adopted with convenience, will be found as profitable, and attended with less trouble than any other. The proportions of it may be varied to suit the wishes or wants of every cultivator.

In

In a field containing four acres, I first grew a crop of turnips, which were clean hoed, and left very free from weeds; they were afterwards fed upon the land, which was immediately (in the beginning of March) manured with ten loads of dung, first ploughed in with a common plough, and afterwards trench-ploughed about fourteen or fifteen inches deep; two acres of which were harrowed very fine, and the feed sown about the middle of March, (though in general I prefer sowing towards the latter end of that month, as I have always found the plants come up nearly as soon as the earlier sown, and attended with fewer weeds.) The carrots came up regularly and well, were ready to hoe in the beginning of May, and tolerably free from weeds; so free indeed that they were hoed out with large hoes, and proved an excellent crop. The other two acres (part of the four which had been turnips) were prepared by ploughing and manuring as for the carrots, and set with potatoes, which came up very clean, and proved an abundant crop. In the succeeding year I grew one acre of carrots (with the former preparations) on the land where the potatoes grew, and one acre of potatoes where the carrots had grown; the other two acres were turnips. Ever since, for eight or ten years, this field has grown turnips,

turnips, carrots, and potatoes, in the rotation above-mentioned: the carrots and potatoes coming upon the same ground only once in three years, the turnips every other year, whereby the land is become so clear of weeds, and so rich, that my crops are annually better, and the expence of hoeing lessened at least one half. The soil, on which this husbandry has been practised, is a good loam, inclined to sandy.

I have kept so few sheep, and observed them so little, that I cannot presume to offer my thoughts to you upon that subject.

Mr. ARTHUR YOUNG's observation, that " the " accounts are so amazingly contradictory, that no- " thing can be gathered from them," is as true as I am inclined to think his conclusion is, " that " moisture is the cause;" and in this opinion I am confirmed as far as a single instance can confirm me. It was in the case of a paddock adjoining to my park, which had for several years caused the rot in most of the sheep which were put into it. In the year 1769, I caused it to be under-drained with covered drains, which have worked well ever since, effectually curing its wetness; and notwithstanding I have since kept my sheep in it, I have never killed one whose liver has been at all affected.

This,

This, surely, seems to prove, as far as a single instance can do it, that by taking away the superabundant water, I have entirely prevented the disorder. Whether the same arose from plants peculiar to a wet soil—whether from the eggs of insects lodged on such plants—whether from the nature and quality of all or any plants growing in such situations—or whether, as some have thought, from the mere act of lodging on such land,—are questions of more curiosity than use. The mere knowledge of the means of preventing so dreadful an evil being sufficient to answer every wish and purpose of the farmer.

Mr. BILLINGSLEY's opinion, confirmed by his experience, of the impropriety of burn-baiting, coincides entirely with the idea I had ever conceived of that practice. It can never be good, but where the soil is very deep, and full of large coarse roots and other vegetable substances. The earth itself, when burnt, I have found to be a mere *caput mortuum.* I am,

Your obliged friend,

J. B.

ARTICLE

Article LIX.

An Abridgement of several Letters published by the Agriculture Society *at Manchester, in consequence of a Premium offered for discovering, by actual Experiment, the cause of the* Curled Disease *in Potatoes.*

LETTER I.

THE writer of this letter is of opinion, that this disease is caused by an insect produced by frost or bad keeping before setting; and that the newest kinds, such as have been raised within these nine or ten years, are most apt to curl, because they will not stand to be kept in winter and spring before setting, as the old kinds will; for in one experiment he took an equal quantity of fine potatoes (what are called Manley's) out of the heap; one part he kept moist and cool, which made them full of virtue, and firm; so that when they came to be set, there was moisture to dissolve the set, and feed the branch, and not one curled amongst them.

The other part he kept dry and free from wet, till wrinkled and soft, and the moisture almost expended, so that when set, instead of putrifying and decaying,

decaying, the set itself received nourishment from the ground, became solid, and harder than ever, and and all in a manner curled.

In autumn 1776, he got up a bed of potatoes to lay by in winter, leaving plenty in the ground as regular as possible; and, before the severity of winter came on, covered part of the bed with straw and pease-haulm, and left the other part of the bed uncovered; that part of the bed which was covered was quite free from curled ones, but the uncovered part produced a great many curled, owing, as the writer says, to frost and severity of the weather.

LETTER II.

THE writer of this letter had about a quarter of an acre of potatoes, well manured with cow and horse-dung, and took the greatest care in picking the fine smooth-skinned potatoes for sets: yet nine out of ten parts were curled. He attributes the cause of this disease to a white grub or insect, which he found near the root, about half an inch long, with eight or ten legs, its head brown and hard; as upon examining a number of the curled roots, he found them all bitten, chiefly from the surface to the root, which of course stopped the progress of the sap, and threw the leaf into a curl.

The

The uncurled roots were not bitten. He tried a few experiments as follows:—First, he put soot to the insects in the rows for two days; and after that, he put lime to them for the same time, but they still kept lively: next he put a little salt, which destroyed them in a few hours. From which he infers, that if coarse salt were put into the ground at the time the land is preparing for potatoes, it would effectually cure this distemper.

LETTER III.

THIS writer attributes the cause of the disease to the method of earthing the stems while in cultivation; and that the branch, striking root into the new earthed-up soil, produces potatoes of such a nature as the year following to cause the disease complained of.

To prevent the disease, he recommends the sets to be taken from those potatoes that have not bred any from the branch covered; or otherwise, to dig the part the sets are to be raised from.

LETTER IV.

THIS writer thinks that the disorder proceeds from potatoes being set in old tilled or worn-out ground; for, though those potatoes may look tolerably

tolerably well, yet their sets will most, if not all, produce curled potatoes; hence he is convinced, that no sets ought to be used from old tilled or couchgrass land; and that, in order to have good sets, they should be procured from land that was purposely fallowed for them; from fresh ley land, where they are not curled; or from ley land that was burnt last spring: Plant them on virgin mould, and your potatoes will have no curled ones amongst them; keep them for winter from any other kind.

To avoid the uncertainty of getting good sets, he recommends crabs to be gathered from potatoes growing this year on fresh land, free from curl, and the next spring to sow them on fresh ley land, and continue to plant their sets on fresh ley land yearly, which he is convinced will prevent the curl.

There are different sorts of curled potatoes, some badly curled, others not so bad: those that are badly curled will not be recovered by planting them on fine fresh ley land; and those that are but little curled may be recovered by planting them on the afore-mentioned land.

Some potatoes will have one good stem, and one curled stem, owing to the set having two eyes: one

end

end of which rots, and will have a good stem; the other end is hard, and will have a curled stem.

Several persons have sown seed in old tilled gardens, in hopes to have potatoes free from the curl; but wherever they planted them next year, they have been curled.

All the good potatoes he saw this year, either on fresh ley land, or on old tilled land, were raised from sets that grew upon fresh ley land last year; and where he has seen curled potatoes, he found, upon enquiry, the potatoe sets grew upon old tilled and worn-out land last year. He gives as a general reason for the disorder, that the land is oftener cropped than it had used to be, much more corn being now raised than formerly.

LETTER V.

IN 1772, this writer planted some potatoes by accident full *nine* inches deep: when taken up, many of the plants were rotted, and a few curled. He kept the whole produce for seed, and planted two acres with it in 1773, not quite six inches deep; the crop was amazingly great; and he did not observe any curled plants among them. In 1774, many of these were planted in different soils, yet they were so infected with the curled disease, that

not

not one in twenty escaped. In 1775, the complaint of this disease became general. In 1776, it occurred to him that the good crop of 1775 was owing to the *accidental deep*-setting of 1772; and that the reason why the *same seed* became curled in 1774, was their being set so near the surface in 1773; he therefore attributes the disease to the practice of *ebb-setting*.

In 1777, he took some potatoes from a crop that was curled the year before, and after cutting the sets, left them in a dry room for a month. Half were planted in ground dug fourteen days before; the other half, having been steeped in a brine made of whitster's ashes for two hours, were also planted in the same land at the same time. The steeped ones came up ten days before the others, and hardly any missed, or were curled. The unsteeped ones generally failed, and those few that came up were mostly curled.

He therefore advises as a remedy,

1*st*. That the potatoes intended for *next* year's sets, be planted nine inches deep.

2*dly*. That they remain in the ground as long as the season will permit.

3*dly*. That

3dly. That these sets be well defended from frost till the beginning of March.

4thly. That the sets be cut a fortnight before planting.

5thly. That they be steeped as above, two hours in brine or lye.

6thly. That the dung be put *over* the sets. And,

7thly. That fresh sets be got every year from sandy soils near the coast, or on the shore.

P. S. At planting, the hard dry sets should be cast aside, for they will probably be curled. Curled potatoes always proceed from sets which do not rot or putrefy in the ground.

LETTER VI.

THIS writer had five drills of the old red potatoes, and four of the winter whites, growing at the same time in the same field. The drills were prepared exactly alike. Among the red not one was curled; the winter whites were nearly all curled. He says he has found, by experience, that the red never curl.

LETTER VII.

TWO of the writer's neighbours had their sets out of one heap of potatoes. They both set with the plough, the one early, and the other late in the season. Most of those early set proved curled, and most of those set late, smooth; the latter on clay hand.

A few roods of land were also planted with small potatoes, which had lain spread on a chamber floor all the winter and spring, till the middle of May. They were soft and withered; yet proved smooth and a good crop. Middle-sized potatoes, withered and soft, which had been kept in a large dry cellar, and the sprouts of which had been broken off three times, produced also a smooth good crop.

Hence he was led to think a superfluity of sap, occasioned by the seed being unripe, might cause the disease. To be satisfied in this, he asked the farmer whether he had set any of the same potatoes this year, and what was the nature of his land. He told him that " he had; that the same pota-
" toes had been set on his farm fourteen years,
" without ever curling; that his soil was a poor
" whitish sand, of little depth; and that he let
" those

"those he designed for keeping grow till they were
"fully ripe."

Hence he concludes, the only sure way to prevent the curl is, to let potatoes, intended for seed, stand till they are fully ripe, and to keep them dry all the winter.

LETTER VIII.

THIS writer set a quantity of the red potatoes, without having a curled one amongst them. His method is, when the sets are cut, to pick out such as are reddest in the inside. On digging them up at Michaelmas, he mixes none of the curled seed among the others. The curled are easily distinguishable, by their stalks withering two months before the rest of the crop.

The cause of the curled disease he attributes to potatoes being of late years produced from *seed* instead of *roots*, as formerly. Such will not stand good more than two or three years, use what method you please.

Last spring, he set the old red and white Russets, and had not a curled potatoe among them.

On

On the lime-stone land about Denbigh, in North Wales, they have no curled potatoes. If this be owing to the nature of that land, perhaps lime might prevent the disease.

LETTER IX.

THIS writer says, that all sorts of grain wear out and turn wild, if sown too long on the same land; the same will hold good in all sorts of pulse, pease, beans, and (as he conceives) potatoes. It generally happens, that those who have most curled potatoes plant very small sets.

Eleven years ago he bought a parcel of fresh sets, of the golden dun kind, and has used them without change to the present year, without any being curled. This he principally attributes to his having always planted good large sets.

About four years since, he thought of changing his sets, as his potatoes were too smooth, too round, and much diminished in size. But the curl at that time beginning to be very alarming, he continued his sets till, part of his crop missing last year, he was obliged to buy new sets this spring, which, being small, were curled like other people's.

He allows, that the curl has frequently happened to perſons who have uſed large potatoes for ſets; for, as all roots are not equally affected, ſome curled ones may be mixed with the reſt.

To prevent the evil, cut your ſets from clear and middle-ſized potatoes, gathered from places as clear of the curl as poſſible; preſerve them as uſual till ſpring. If any are harder, or graſh more in cutting than uſual, caſt them aſide. He would alſo recommend the raiſing a freſh ſort from the crab produced on the ſorts leaſt affected, which in Lancaſhire are the *long-duns*.

N. B. Here follow three certificates from perſons who have raiſed their crops from large ſets of the long-duns, for many years, without being affected with the diſeaſe.

LETTER X.

SET Potatoes with the ſprits broke off, and they will (ſays the writer of this letter) be curled ones; if ſet with the ſprits on, they will not be curled. Again, take a potatoe which is ſprit, and cut a ſet off with two ſights; break one ſprit off, and let the other ſtay on, and ſet it; the former will be curled, and the latter will not.

When

When you have holed your potatoes, take them out before they are fprit, and lay them dry until you have fet or fown them, and you will have no curled potatoes.

LETTER XI.

THIS writer was at the expence of procuring fets at fifty miles diftance, and where this difeafe was not known; the firft year's trial was fuccefsful; the year following he procured fets from the fame place; but one-fifth of his crop was infected. By way of experiment, he planted fets from roots which had been infected the year before, and fome of thefe produced healthy plants, free from all infection.

As every effect muft have a caufe, he fuppofed it might be fome infect, which, living on the leaves, gave them that curled and fickly appearance, as is the cafe in the leaves of many fhrubs and trees. But whether the infect be lodged in the *old fets*, and to be deftroyed at the time of planting, or, proceeding from fome external caufe, can only be deftroyed afterwards, he is not yet certain, although he has made the following experiments:———.

On a piece of ground that had not been dug for twenty years, he planted four rows of fets, which he knew to be perfectly clear: the drills were two feet diftant; the fets one foot diftant in each drill. He then planted on the fame ground four rows with fets from curled potatoes, at equal diftances; in each row were about twenty fets.

Lot 1ft, the curled ftate.

No. 1. without manure, | No. 3. in foot,
    2. in falt, | 4. in quick lime.

Lot 2d, the clear fets.

No. 1. without manure, | No. 3. in foot,
    2. in falt, | 4. in quick lime.

Thofe planted in falt and foot in both lots, were deftroyed. In Lot 1, No. 1 and 4, all curled. Lot 2, No. 1 and 4, quite clear.

This experiment was made on a fuppofition that the infect lodged in the fet, and muft be deftroyed on planting. But of that he is not fully fatisfied. He repeated falt, foot, and quick lime, on the branches of feveral curled potatoes. Salt deftroyed all he touched with it. Lime and foot had, he thought, a partial effect on the plants.

After

After some time, they appeared almost as healthy as the rest. Thus, although he had done little towards the cure, he flatters himself he has pointed out the cause, the insects on the curled plants being not only very numerous, but visible to the naked eye.

LETTER XII.

THIS writer ascribes the cause of the curled disease in potatoes to the frost and bad keeping in winter and spring before setting. They are liable to be damaged by frost after they are set, but this may be prevented by covering. If it be asked why frost did not injure them formerly, he answers, it is only the *new* kinds which are apt to curl. To this may be added, that less care is now taken of the seed than formerly.——To prevent the latter, let them remain in the ground covered with haulm or litter, till the time they are wanted for setting; and, in case no frost touches them afterwards, they will be free from the disease.

LETTER XIII.

THIS writer says, the red potatoe was as generally planted as the winter white and the Lincolnshire kidney are now. The first, being a later potatoe,

potatoe, did not sprout so early as the others. The *white* sprout very early, and therefore should first be moved out of the place where they have been preserved in the winter. Instead of that, they are often let remain till their roots and sprouts are matted together.

On separating them, these sprouts are generally rubbed off, and they are laid by till the ground is ready; during which interval they sprout a second time: but these second sprouts, being weak and languid, will shrink, sicken, and die, and the fruit at the roots will be small, hard, ill-shaped, and of a brown colour.

Now, if putting off the sprouts once or more, before the sets are put in the ground, be the cause, (as he verily believes it is) of the curled disease, an easy remedy is at hand. When the potatoes intended for sets are dug up, lay them in a west aspect as dry as possible: in such a situation they will not sprout so soon.

The best time for removing most sorts is, the first fine day after the 24th of February. Cut them into sets as soon as possible, and let them remain covered with dry sand till the ground is prepared, which should be a winter-fallow. Lay the

the fets in without breaking off any of the fprouts, for the fecond will not be fo vigorous.

This accounts for *one* fprout out of three from the fame fet being curled. The two ftems not curled rofe from two later eyes, and were firft fprouts. The fprout curled was a fecond, the firft having been rubbed off.

LETTER XIV.

THIS writer fays, that laft fpring one of his neighbours cut and fet, in the ufual way of drilling, fome loads of the largeft potatoes he could procure; and more than half of them proved curled. Being a few fets fhort of the quantity wanted, he planted fome very fmall potatoes, which he had laid by for the pigs. Thefe being fully ripe and folid, there was not a curled plant among them.

He apprehends, the others being curled was owing to their not being fully ripe. A crop of potatoes, fet this year in rows, on ground that had borne a crop of them laft year, were moftly curled; but many plants came up from feed left in the ground laft feafon, and there was not a curled one among them.

LETTER

LETTER XV.

ALTHOUGH, the writer of this letter obferves, it is generally believed, that taking up potatoes, intended for the next year's fet, too foon, is a *principal* caufe of the curled difeafe, he has his doubts refpecting it; having let fome remain in the ground all winter, and vegetate the following fpring, fome of them were curled, and others not, in like manner as thofe fets proved which he took up and planted as ufual. This, therefore, he infers, cannot be the principal caufe. The old white rough, and the kidney potatoes, are as fubject to this difeafe as the reft. Red potatoes of moft kinds are feldom affected with it.

LETTER XVI.

OF late years, this writer fays, great improvements have been made in fetting potatoes, and cutting the fets. The ground is dreffed cleaner and dunged ftronger. Many people in drilling wrap up the fets entirely in the dung; by which means, though their potatoes are larger, the difeafe feems to be increafed. They alfo cut their fets out of the richeft and largeft potatoes, which is perhaps another caufe of this evil. In cold countries, where they fet their own feed, which has

grown

grown on poor land, with lefs dung, they have no curled plants. On the contrary, when they bought rich and large potatoes for feed, they have been curled in great quantities. He believes, the richnefs and largenefs of the feed to be the caufe of the evil; for he does not remember to have feen a curled ftem which did not fpring from a fet of a large potatoe.

LETTER XVII.

THIS writer apprehends the curled difeafe in potatoes to proceed from a defect in the *planta feminalis* or feed-plant; and from comparing curled ones with others, there appeared to be a want of, or inability in, the powers of expanding or unfolding the parts of the former; which, from this defect, forms fhrivelled, ftarved, curled ftems. On examining fome of the fets, at the time of getting in the crop, he found them hard and undecayed; fo hard indeed, that fome of them would not be foft with long boiling. This led him to think that fome manures might have the fame effect on them as tanner's ooze has on leather, and fo harden them that the embryo plant could not come forth with eafe; but a clofer examination taught him otherwife, and that they grow equally in all manures.

Some

Some have thought that the fermentation is occasioned by too great quantities being heaped together; but the writer has seen an instance, wherein a single potatoe, preserved by itself, when set, produced stems of the curled kind. He thinks the most consistent and rational opinion is, that the disease is occasioned by the potatoes being taken from the ground before the stamen, or miniature-plant, is properly matured and ripened.

For let it be observed, that the potatoe, being a native of a warmer climate, has there more sun, and a longer continuance in the ground, than in its present exotic state; consequently, it has not the same natural causes *here* to mature the seed plant, as in its native state. We ought, therefore, to give all the opportunities our climate will admit for nature to complete her work, and fit the *stamen* for the next state of vegetation, especially in those intended for seed. But if the potatoe be taken up before the seed-plant be fully matured, or the air and sap vessels have acquired a proper degree of firmness or hardness, it must, when thus robbed of further nutrition, shrivel up; and when the vessels, in this immature state, come to act again in the second state of vegetation, they may produce plants which are curled.

If

If it be asked, why are they more common now than formerly? he answers, that before the present mode of setting them took place, people covered them, while in the ground, with straw, to protect them from frost.

If it be asked, why one set produces both curled and smooth stems? he answers, we suppose every eye to contain a *planta seminalis*; that all the embryos, or seed-plants, contained in one potatoe, are nourished by one root; and that, as in ears of corn, some of these seed-plants may be nourished before others.

One of his neighbours, last year, set two rows of potatoes, which proving all curled, he did not take them up; and this year there is not a curled one among them. Such potatoes, therefore, as are designed for seed, should be preserved as long in the ground as possible.

LETTER XVIII.

THIS writer advises such sets to be planted as grow in moss-land; and, he says, there will not be a single curled one the first year. This is affirmed by the inhabitants of two townships, where they grow amazing quantities.

A medical

A medical gentleman sowed last year two bushels of sets from one of the above places, and had not one curled; but on sowing them again this year he had a few.

N. B. Although the foregoing letters do not point out with certainty the real or general cause of the curled disease in potatoes, or discover any specific remedy which reaches all cases, yet as they contain many interesting observations both on the disease itself, and the best methods hitherto adopted for preventing it, we think they are not improperly introduced in this work. And, notwithstanding there seems to be a diversity of opinions in the writers, occasioned by the different appearances of their crops, and the seemingly contrary effects of the means used to prevent or cure the disease, we conceive, that the following *general propositions* may be fairly drawn from the whole:——

1*st*. That some kinds of potatoes are in the general much more liable to be affected by the disease than others; and that the Old Red, the Golden Dun, and the Long Dun, are the most free from it.

2*dly*. That

2*dly*. That the difease is occasioned by one or more of the following causes, either singly or combined; 1st, by frost, either before or after the sets are planted; 2dly, from planting sets cut out of large unripe potatoes; 3dly, from planting too near the surface, and in old worn-out ground; 4thly, from the first shoots of the sets being broken off before planting, by which means there is an incapacity in the *planta seminalis* to send forth others sufficiently vigorous to expand so fully as they ought.

3*dly*. That the most succesful methods of preventing the difease are, cutting the sets from smooth middle-sized potatoes, that were fully ripe, and had been kept dry after they were taken out of the ground; and without rubbing off their first shoots, planting them pretty deep in fresh earth, with a mixture of quick-lime, or on limestone land.

ARTICLE

Article LX.

Description of, and Observations on, the Cock-Chaffer, *in its Grub and Beetle States.*

[By the Secretary of the Society.]

AS there are few insects more prejudicial to the farmer than that generally known by the name of the *Cock-Chaffer*, I beg leave to make a few observations thereon.

In different parts of this kingdom these insects are called by different names, such as the *Chaffer*, the *Cock-Chaffer*, the *Jeffry-Cock*, the *May-bug*, and (in Norfolk) the *Dor*.

In what class Linnæus ranks them, I do not remember; but they seem to be the *Scarabeus arboreus vulgaris major* of Ray.

When full grown in their grub-state, they are near an inch and a half long, and as big as a child's little finger. Their heads are red, their bodies soft, white, and shining, with a few hairs on the back. They have three hairy legs on each side, all placed near the head, in which are two forceps or jaws, like the hornet; with these they cut asunder the

the roots of grafs, corn, &c. and frequently deſtroy whole fields in a ſhort time. In this *eruca* or grub ſtate, they continue three and ſometimes four years.

In their beetle-ſtate they have two pair of wings; the one filmy, and the other ſcaly. The *interior* pair are folded up in a curious manner, and remain hid, unleſs when expanded for flight. The *elytra*, or caſe-wings, are of a reddiſh brown colour, and ſprinkled over with a fine white powder, like the auricula. The legs and tail (which is pointed) are whitiſh. The body is brown, except at each joint on the ſides of the belly, which is indented with white. The circles round the eyes are yellowiſh; the antena ſhort, and terminated by fine lamellated ſpreading tufts, which the creature expands more or leſs as it is briſk and lively or otherwiſe.

The firſt account I find of theſe deſtructive inſects, is given by *Mouffett*, who tells us, that in the year 1574, ſuch a multitude of them fell into the Severn, that they clogged, and even ſtopped, the wheels of the water-mills.

There is alſo an account in the Tranſactions of the Dublin Society, that the country people ſuffered

ſo

so much in one county, by the devastation these insects made, that they set fire to a wood several miles in length, to prevent their further progress.

In the day-time they seldom fly about, but conceal themselves beneath the leaves of oak, sycamore, maple, hazel, lime, and some other trees, which they soon eat to a skeleton; but about sun-set they are all on the wing, and fly about the trees and hedges as thick as a swarm of bees.

While in their grub-state, they entirely destroy all the grass, corn, or turnips, where they harbour.

I have seen fine meadows withered in May and June, and as brown as thatch.

These grubs generally lie near two inches below the surface, and eat the roots of the grass so regularly, that I have rolled up many yards of the withered turf as easy as though it had been cut for a garden.

When they attack turnips, they eat only the middle of the small root; but by that means kill all they bite without remedy.

Neither

Neither the severest frosts in our climate, nor even keeping them in water, will kill them. I have kept some in water near a week; they appeared motionless; but on exposing them to the sun and air a few hours, they recovered, and were as lively as ever. Hence, it is evident, they can live without air. On examining them with a microscope, I could never discover any organs for respiration, or perceive any pulsation.

Hogs will root up the land for them, and at first eat them greedily; but seldom meddle with them a second time. To rooks and crows they seem to be a high regale. When numerous, they are not destroyed without great difficulty; the best method is, to plough up the land in thin furrows, and employ children to pick them up in baskets: and then strew salt and quick-lime, and harrow in.

About thirty years since, I remember many farmers' crops in Norfolk were almost ruined by them in their grub-state; and in the next season, when they took wing, the trees and hedge-rows in many parishes were stript bare of their leaves as in winter. At first the people used to brush them down with poles, and then sweep them up and burn them. One farmer made oath, that he gathered eighty bushels;

bushels; but their number seemed not much lessened, except just in his own fields.

Their mode of *coupling* is singular; and the time of their continuance in that act, sometimes two or three days. I have seen one of them fly in that state, with the other hanging pendant from its tail; and am in some doubt whether (like snails) they are not *hermaphrodites*, as there seems to be mutual insertion.

They deposit their eggs in the earth. The first year the grubs are very small, and do little mischief; the second year they are increased to the size of a goose-quill, and are very injurious to the herbage; the third year they attain full size, and fly.

E. RACK.

Bath, March 26, 1780.

APPENDIX.

APPENDIX.

APPENDIX.

A PROPOSAL

FOR THE

FURTHER IMPROVEMENT

OF

AGRICULTURE.

BY

THE REV. WILLIAM LAMPORT,

THIRD EDITION.

Multum adhuc reſtat operis, multumque reſtabit; nec ulli nato poſt mille ſecula precluditur occaſio aliquid adjiciendi.

PLIN. HIST. NAT.

MDCCLXXXVIII.

TO THE

SOCIETY OF ARTS, MANUFACTURES, AND

COMMERCE, IN LONDON,

THE

AGRICULTURE SOCIETY AT BATH;

AND

THE OTHER AGRICULTURE SOCIETIES

IN

GREAT-BRITAIN AND IRELAND;

THE FOLLOWING PROPOSAL IS ADDRESSED,

WITH

ALL DUE DEFERENCE AND RESPECT,

BY THEIR

MOST OBEDIENT SERVANT,

THE AUTHOR.

PREFACE.

THE only disagreeable circumstance which the Author experienced in drawing up the following Proposal, was the necessity he found himself under of enlarging on the prejudices and untractableness of illiterate Farmers and their servants.

Censure can be no pleasing task, except to those who deserve the severest censure themselves,—the proud, the envious, and the malicious.

But the Author's intention is to raise the humble spirit of Agriculture, and to convince those who are practically employed in it;

it, that the more it is made an object of reason, the higher it will rise in the scale of perfection.

He asserts nothing upon the report of others: his own reason, aided by some practice, has enabled him to exhibit the following Proposal for the improvement of an Art, on which the wealth, strength, and prosperity, of this nation principally depend.

A PROPOSAL

A PROPOSAL FOR THE FURTHER IMPROVEMENT OF AGRICULTURE.

A Noble spirit, for making improvements in Agriculture, hath lately gone through this nation, for which posterity will thank the present age in terms of the highest approbation. The principles on which those improvements have been conducted are as judicious as the subject is important; and it is highly probable, that many good effects will take place in every part of the kingdom.

Agriculture has been considered of national importance by the most discerning part of mankind in all ages.

Every civilized nation, at one period or other, have been convinced of its intrinsic excellence; and the wisest men, of every age and country, have unitedly bestowed the highest encomiums on it. In the present times we have the satisfaction of

seeing,

seeing, that the noble, the wife, and the learned, do not think it beneath them to rescue it from that obscurity in which it had long been involved, and to bring it forward to public view, under the sanction of their own practice.

But it is not the purpose of this Essay to write an eulogium on the dignity and utility of Husbandry, either by adverting to the dispensations of God Himself towards the Jews,* or by extracting from the writings of the most eminent men, ancient and modern. This is needless. Rather let us collect some of their best ideas concerning the means of advancing Agriculture to the highest perfection, and thereby fulfil, if possible, the purpose of this essay, which, it is hoped, will recommend itself to the attention and regard of the public, merely from the importance of the subject.

---

* Vide S. S. passim, particularly Lev. 25. The command in this chapter, that every seventh year should be a year of rest, or fallow, to the land; and that the produce of the sixth year should supply the nation for three years, had a peculiar tendency to make the Jews *skilful*, as well as industrious, in works of husbandry; and, I believe, it is pretty well known to every skilful cultivator, that land well tilled, dressed with proper manures, and sown or planted with a judicious rotation of crops, will scarcely ever stand in need of a fallow, till the seventh year at least; and that the labour of the sixth year will be peculiarly blessed to such an husbandman. This much, however, is certain, that ground, cultivated as above, will frequently resist the ill effects of intemperate seasons, by which neighbouring fields greatly suffer when under unskilful and indolent management.

JULIUS

Julius Cæsar,[*] speaking of the manners of the *Germans* in their rude uncultivated state, makes the following instructive observations, which are applicable indeed to all people in similar circumstances:———

' Agriculture they disregard; their diet consist-
' ing chiefly in milk, cheese, and flesh: for none
' of them have any certain quantity of ground,
' or even country, which they can call their own.
' But their magistrates and chiefs allot, for one
' year only, among the scattered inhabitants[†] and
' their tribes who associate together, such a por-
' tion of land, and in such a district, as they think
' proper; and then oblige them to reside at some
' other place for another year. They assign seve-
' ral reasons for this conduct:—That the people
' might not be induced to exchange the study of
' war for that of husbandry;—that they might
' not wish to increase their settlements, and so the
' stronger expel the weaker from their possessions;
' that they might not erect any buildings, except
' barely to keep out heat and cold; &c.'.

[*] De Bell. Gall. lib. vi. cap. 22.

[†] Gentibus. On this word, see a judicious criticism of the Monthly Reviewers, in their account of Holdsworth's remarks on Virgil, for June 1768, p. 426.

A country

A country will be cultivated only in proportion as its inhabitants advance in civilization. Nations will not begin to civilize themselves, till they cease migrating from place to place: neither will a man attempt to cultivate any spot, 'till he can say THIS IS MINE. But when men unite together for mutual protection and advantage, and settle in one place, the cultivation of that spot immediately becomes necessary, that it may supply them with the conveniences of life. Property, therefore, must be gained and defined, settled and secured. These are circumstances on which the advancement, if not the very existence, of Agriculture depends.

But these are not all. There are two others of equal importance to its improvement and prosperity: the one is, the *fruit* of a man's labour must be secured to him: the other, that as the wants of men increase in consequence of civilization, the earth must be encouraged to yield proportionable supplies.

This, however, can be effected only so far as the powers of the human mind are enlarged in consequence of civilization. Husbandry can rise no higher than the knowledge of those who are engaged in it will permit. It hath been indebted
for

for its principal improvements, not only to the *natural* abilities of the cultivator, but to an *education* formed upon an acquaintance with other branches of science.

Whenever any of the above circumstances fail, Agriculture must feel a stagnation :—in proportion as they are regarded, will be the progress made in it, and its success.

This appears to be the case in fact; for these circumstances, especially the last of them, were not heretofore sufficiently attended to by this nation; which will fully account for the defective state of husbandry in former times, its slow progress, and its present improvements; while *it also points out the most probable method of carrying it still nearer to perfection.*

If we expect to find Agriculture in a thriving state before the Reformation, we shall be disappointed : it was indeed considered of importance; but *the fruit of a man's labour was not secured to him; and the nation was immersed in gross ignorance.* The feudal constitution, the military disposition of the people, and the tyranny of popish ecclesiastics, were unfriendly to skilful and vigorous cultivation.

There

There was no great encouragement for the owners of estates to exert themselves in the cultivation of them, while others were to reap the fruit of their labours: this was, therefore, left to their meaner vassals, whose spirits were sufficiently humbled to submit to almost any imposition. The same reason which is to be given for the uncultivated state of Italy, though in itself the garden of the world, may be assigned for the general disgrace into which rural œconomics had fallen in England, till the time of the Reformation.—" For one may venture to pronounce, without prejudice, that Agriculture, *cæteris paribus*, will always flourish most in free governments and protestant countries."*

In such a situation of things, when Agriculture was, as it were, banished into desarts, and in every respect took up its residence among mountains and vales, where knowledge had made small progress;—when the mind of the peasant was not enlightened by the rays of science; when he tilled the earth merely by the labour of his hands and the sweat of his brow, without any fixed principles; it is not to be supposed, that any considerable improvement could be made by him.

* Harte's Essay I. p. 67.

Nor

Nor was this all. Admitting that the principles of vegetation had been accurately delineated to his view, or experiments founded thereon propofed, it was not for him to inveftigate the one, or practife the other, while ecclefiaftical tyranny prevailed, and he knew that the priefthood would reap the far greater part of the fruits refulting from his labour.

Tyranny over the mind will ever retard the progrefs of every kind of knowledge.

But even after the Reformation, although many of the arts and fciences were cultivated with peculiar fpirit, Agriculture did not receive encouragement proportioned to its great importance. Every thing cannot be attended to at one time.

A new world had been difcovered, which opened the brighteft profpects to thefe kingdoms, and the attention of England was fixed chiefly on trade and commerce. This circumftance, which for a while appeared to be a principal impediment to hufbandry, and was the caufe of little attention being paid to FITZHERBERT,[*] proved in the event

[*] The father of Englifh Hufbandry: made Judge of the Common-Pleas about the year 1524. His book of hufbandry was printed in 1534, after forty years' attention to the fubject, in his receffes between the Terms.

one of its principal promoters. By commerce, the various productions of different parts of the earth have been brought into this kingdom, and intrusted to the care of the skilful botanist and gardener; who, having naturalized them to this climate, commit them to the care of the husbandman. In return, Agriculture has ever since been assisting commerce in the increase of corn, hemp, flax, madder, &c. &c.; and in proportion as both have been attended to, it is evident they have mutually assisted each other.

But as improvements prevailed, the importance of husbandry, in a national view, became daily more and more conspicuous;—the disadvantages and impediments it met with, under the management of common farmers, began likewise to appear. The weeds sprang up with the wheat, and skill was wanting to prevent this evil.

To check these weeds, by enlarging the views of those intended for the profession of Agriculture, was the noble attempt of the great MILTON, who not only recommended, but established, a school, in which rural œconomics were to bear a principal part in his system of education. His pupils were to read the works of Cato, Varro, Columella, &c.

&c. on Agriculture.* But unhappily, his loss of sight prevented him from realizing in practice what he had so judiciously adopted in theory.

That EVELYN, one of the most useful men of the age in which he lived, entertained the same sentiments as Milton, appears from the preface to his Sylva.† To him the nation is now, and will be for many years to come, greatly indebted for the strength of her navy.

To form a glorious triumvirate, we can invite the very modest and sensible Mr. COWLEY, in support of the same plan. He recommended a College to be erected in each University, and the appointment of professors for the instructing of young persons in the principles and *practice* of this useful employment.‡

But, as it is always the fate of the most useful designs to meet with difficulties at the beginning, Agriculture itself began, soon after, to fall from its

---

* Letter to Hartlib, and Biog. Britan.

† Many millions of timber-trees (besides infinite others) have been propagated and planted at the instigation and by the direction of this work.——See Dedication of Sylva to Cha. II.

‡ Cowley's Works, vol. ii. p. 656, 7.

flourishing state into national disregard. This, however, is easily accounted for.

I have laid it down as a general rule, that civilization encourages husbandry: yet it is possible that rural œconomics may be impeded by this very civilization, unless it be well regulated. A nation may be civilized to so high a degree of refinement, as that the politer part of its inhabitants will associate in cities and towns, and attend to nothing but pleasure and the fine arts. The consequence is, that Agriculture will be nearly in the same predicament as it was before the commencement of civilization.

In such a state of *false* refinement, the cultivation of land will be considered as beneath the notice of the rich and the learned, and be left to the ruder part of the people.

Such was the state of this nation in the reign of that gay Prince Charles II.; and could any thing else be expected but that Agriculture must severely suffer, in an age so deeply immersed in luxury, pride, and dissipation? especially if it be considered that the persons who paid the closest attention to it, had " crept into the confiscated estates of the no-
" bility,

" bility, gentry, and clergy," and were many of them originally in very inferior stations.

At that period, the maxims of the celebrated BACON, the example of MILTON, the efforts of the ROYAL SOCIETY, the proposal of COWLEY, the complaint of EVELYN, and his just observations on the necessity of an enlarged education, in order to improve the lands of England, were exhibited in vain.

It was to little purpose that the ministry, after the Restoration, permitted the exportation of wheat;* it increased *tillage*, but did not improve *the mode of culture*, or reconcile the nobility and gentry " to what had been the object and care of " mean and despised persons."

Thus Agriculture fell into disrepute, and was driven back again to the mountains and vales, where FITZHERBERT first found her; with this difference only in her circumstances, that she might be more easily recalled by the writings which were extant, whenever the nation should be restored to its characteristical sedateness.

* Combrune's History of the Prices of Corn, Ann. 1663 and 1670.

Whenever any scheme of real utility and national importance is formed by men of genius and true patriotism, the worst kind of impediment it can meet with is, that of national supineness and inattention. If it be not actually opposed, it is not promoted; and if people do not reflect on it, they cannot see its importance.

Nothing, however, can totally check the vigour of great minds. EVELYN, in the midst of this general indifference, published, in the year 1675, his *Terra,* or a *Philosophical Discourse on Earth,** which, with the assistance of former publications, began to open the eyes of his countrymen to their true interest, to the dignity of his subject, and the necessity of *more than a superficial knowledge,* in order to make improvements in it.

The next writer we shall mention is, Lord MOLESWORTH, who, in his *Considerations for the promoting of Agriculture, and employing the poor,* makes the following judicious remarks, quite in point to the purport of this essay. " As to Agriculture, I would humbly propose that a *school for husbandry* should be established in every county, wherein a master well skilled in Agriculture should teach at

* In 1778, Dr. A. Hunter republished this work with notes.

a fixed

a fixed yearly falary: and that TUSSER's old book of hufbandry fhould be taught the boys to read, to copy, and get by heart; for which purpofe it might be reprinted."*

Complaints of the impracticability of illiterate peafants making any confiderable improvements in rural œconomics, and the neceffity of affifting them, began now to be as general as juft; being founded on facts and fad experience, which were pregnant with many pernicious confequences. It was clearly feen, that they could not deviate from the beaten track; that they were not capable of reflecting on the nourifhment of plants, in order to increafe vegetable food by judicious and frequent ploughings and fuitable manures; of introducing new claffes of vegetables, however advantageous; or of making any experiments on fcientific principles; efpecially as they knew, that if *thefe* failed, they fhould rifque the failure of their rent. On all accounts, therefore, they muft continue in the courfe marked out by their anceftors, however defective and injudicious.

Thefe imperfections having been long obferved and lamented, feveral gentlemen of publick fpirit

* Dublin, Ann. 1723, Harte's Effay I. p. 156.

(the

(the leader of whom was the famous Tull) took their estates into their own hands, and cultivated them with—' spirit, taste, and sense,'—by regulating the course of crops according to the nature of the soil—by banishing wasteful fallows*—by destroying weeds—by stirring and pulverizing the earth while a crop was growing, and thereby preparing it for the immediate reception of a succeeding one—by introducing new plants for the better support of man and beast in winter as well as summer, &c. But, unhappily, these capital improvements remained for a long time within the circle of those farms where they originated, or those counties where such public-spirited gentlemen had set the example by their own practice. These modes of cultivation were novel; on *this* account they were slighted, if not derided, by the generality of the common farmers. The principles on which such culture was founded were above their comprehension; it must therefore necessarily be, as they fancied it, too expensive for them to run the risque of practising.

* We cannot fully coincide with Mr. Tull, in the idea that *all* fallows are *wasteful*. We readily grant, that by a judicious succession of crops, and ploughings often repeated, the annual quantity of fallow-ground might be greatly reduced, without impoverishing the soil; but we still think that *some* fallows are annually necessary, especially where the land is naturally poor.

This

This circumstance gave rise to another plan, in itself most honourable and benevolent, namely, the establishment of a Society in London for the encouragement of Agriculture, &c. who, by bestowing large premiums for the greatest crops on given quantities of ground, effectually secured the farmer under any risque he might run. It was naturally imagined this would have answered the end proposed. But if we may determine from Mr. BAILY's register of the persons to whom premiums have been adjudged, most of the candidates have been far above the rank of common farmers.

The diffusive plan adopted by that illustrious Society was intended to include every farmer; but we find it has attracted the notice of very few, except the more civilized part of them; while many parishes, I was going to say almost whole counties, at a distance from the capital, remain uninterested about every thing relative to the Society, if not totally ignorant of its existence.

However, the advantages arising from that excellent institution excited and established others of a similar nature, in counties remote from the metropolis; each of which hath thrown additional light on the subjects of Agriculture. Many experiments

riments and new discoveries* have been made, all concurring to prove that very little of the true principles of vegetation was understood by those

* Here let me congratulate the public, particularly, on the invention of the *Oil Compost*, by the ingenious Dr. Hunter, of York, which, having tried myself both on wheat and turnips, growing on *very poor* ground, I found it answer even to admiration, particularly with regard to the latter. I think it one of the best preservatives from the Fly; and can, therefore, recommend it as a valuable acquisition, being likewise a good manure, where dung is scarce, or the carriage of it expensive. It might be of signal service in the improvement of waste ground till dung can be raised: for were such ground to be pared, the turf dried and thrown into heaps for burning, and, in the mean time, the earth between the heaps ploughed carefully and to a proper depth; and, after this, the ashes spread, ploughed in, and well harrowed, the turnip-seed sown, and the oil compost spread by hand at the same time, or just on the appearance of the turnips, I think there would be but little danger of a good crop, if well hoed; as ashes are peculiarly favourable to the growth of turnips; and if they were to be eaten off with sheep, and the next course barley or oats, with grass-seeds suitable to the soil, I am persuaded this would prove one of the most speedy, effectual, and cheap methods of improving such waste lands; especially, if in a dry spring the ashes were to be spread the latter end of April, and the land sown with buck-wheat, to be ploughed in as a manure when in blossom, and then the turnips to be sown, the oil compost spread, and both to be harrowed in as above. I have raised pretty good turnips after buck-wheat ploughed in, without the oil compost; but as the expence is not great, the using of both would perhaps be the *ne plus ultra* of improvement in the circumstances above-mentioned. The grand fault committed after paring and burning is, taking too many exhausting crops.

A coarser kind of oil, I am informed, is made in Cornwall, from pilchards, which would, in all probability, make the compost come considerably cheaper.

For the method of making the compost, see Dr. Hunter's Georgical Essays, vol. I.

who had undertaken to supply the nation either with food, or materials for carrying on the linen or woollen manufactures.

Who are the persons that have so increased the produce of wheat, that the London Society admit no claims for the premium under five quarters per acre; although the average quantity, produced the last very favourable year, [1781] did not, in all probability, far exceed three quarters per acre, throughout the kingdom,* if you except the estates of gentlemen?

Who are the persons that have so much promoted the growth of cabbages, carrots, madder, &c. by field culture? Who have been and are likely to be, in general, candidates for the premiums of the above laudable Society? And who are they that are attending to husbandry according to its true principles?—Not the uninstructed farmers, who are yet but little acquainted with the subject as to several of its essential and fundamental points. They are ignorant of the various properties of different manures, and how they respectively operate, particularly on different soils; nay, on the same soils when differently circumstanced; as well as

* In this, we believe, the Author is mistaken.

the different *inherent* qualities of foils apparently fimilar, &c. But without the knowledge of thefe properties and qualities, miftakes have been and will be committed by farmers, and difappointments in their crops will happen, which they know not how to account for, nor in what manner to prevent in future. A remarkable inftance of the neceffity of making experiments on the different qualities of foils, may be feen in the Complete Englifh Farmer, p. 104, 5.

Were thefe things better underftood, they would not continually manure their ground with dung, where lime and marle are eafily procured, nor conftantly *repeat* lime on the fame field, becaufe it carried feveral good crops while abounding with vegetable food: they would not take three exhaufting crops in fucceffion, nor proceed in the fame courfe of crops on every kind of foil.

Thefe miftaken notions and practices cannot be removed merely by the diftributions of premiums. The ideas of illiterate farmers will not be much rectified by many of our publications on hufbandry, which fome cannot, and the generality are too opiniated to read: Add to this, that whoever implicitly follows the theory contained therein, will
often

often be led into errors that would end in lofs and difappointments.

Premiums have a tendency to excite a fpirit of emulation and induftry, to increafe the produce of the earth, according to the different mode to which any diftrict or county hath been accuftomed; but a common farmer, fhould he become a candidate, will have no more chance of fucceeding againft perfons of a liberal and extenfive acquaintance with the principles and practice of Agriculture, than any one of his draft horfes could have in attempting to keep pace with his landlord's hunter.

Can the bare donation of premiums give inftruction to the mind? Ought not this to be communicated in youth, when the difpofition is docile? Enlarge the views by cultivating the underftandings of young perfons while they are moft fufceptible of impreffions, and free from prejudices, and they will be continually increafing in knowledge as they grow in years; but if the mind be not improved early, the confequence will be, in general, (for the exceptions are but very few) that they will pertinaciously adhere to old cuftoms, however abfurd.

Whoever

Whoever hath been much conversant with the common farmers, (and it is by *them chiefly* that our lands are cultivated) must have observed that they generally associate together, communicate their ideas to each other in their own way, gaining no more information from one another than the knowledge each hath obtained, can bestow; and that their observations are founded on *their own customs* in the country where they reside. They are a class of people *sui generis*, and stand at a distance, as it were, from a man of learning; and unless he can make himself very familiar with them, and converse in their own stile, it is most probable that they will either entirely mistake his meaning, or inwardly sneer at some expressions which they do not understand: and thus go away unimproved as they came, or resolved not to follow his advice. Of great importance, therefore, is Education, to extend and call forth the powers of the mind, and to render it ductile and teachable!

Therefore, until Agriculture is erected on this enlarged basis, will it not continue a vague and abstruse study in itself, and remain far short of that degree of perfection, which our public-spirited and useful Societies would wish to see it attain?

If

If Agriculture is to be improved by learning, why should not this class of people, the Farmers, be better educated? They are capable of improvement. Let them be well instructed; and improvements in husbandry will soon make their way into every village, perhaps without much assistance from premiums. However, when instruction is stimulated by premiums, the great end, I trust, will be still more effectually answered.

Every one who reflects justly must be sensible, that it is with Agriculture as with physic. While facts and experiments are producing and increasing the best knowledge, it is necessary that those who may hereafter engage in either of the professions, be instructed in the first principles of the one, and the practice of the other.

Agriculture is a science as well as an art; and some general scientific knowledge is requisite before that art can be practised with any rational hope of full success; unless quacks may be allowed to perform perfectly well in Agriculture, although they are continually breaking the sixth commandment in physick.

Mr. YOUNG indeed observes, that "experience is an admirable foundation for any kind of structure;
but

but in Agriculture she must be the structure itself, not the foundation."*

But I would have taken the liberty to ask, What is to be the foundation of this structure? had not Mr. Young himself pointed it out, when he ingenuously confesses, " in many instances I have been a very bad farmer, and acted contrary to the dictates of good husbandry."†

No one will pretend to deny that experiments are the life and soul of husbandry,—but they must not be made at random; for to what can such experiments tend, except to the frequent disappointment of the farmer, and to the publick loss?

Indeed the encomium which Mr. Young hath so justly passed on Dr. Home, evidently proves, that the practical part of Agriculture must receive considerable benefit from scientifick knowledge.

It is hoped that these remarks will not be considered as censure on Mr. Young, or ' as a cavil at excellence.' They are intended only to place this subject in the most enlarged point of view.

* Experimental Agriculture, pref. p. 15.   † Ibid. p. 6.

However

However short and defective the above account of the state of Agriculture in this kingdom at different periods may be; yet I hope I have made it appear—that it is much indebted for its present improvements to learning and civilization—that whatever deficiences it still labours under, they are owing to a defect in the education of farmers in general—that it hath a close connexion with other branches of science—that learning and experiments must go hand in hand—that the proposals of those sensible and learned men above quoted, for establishing schools of Agriculture, were founded on enlarged views, substantial grounds, and the greatest propriety—and that the little attention which has been paid thereto can be attributed to nothing else but certain temporary circumstances, which retard improvements of one kind or other in every age.

Agricultural Societies were not established when those gentlemen wrote: and it can hardly be supposed that, whatever propriety or utility there might have been in their plan, they alone could suddenly turn the regard of the nation to a subject of which it had then scarce any idea.——The case is now otherwise. Agriculture hath arisen, like a star of the first magnitude, in our hemisphere; and many of the wise men of our nation, of all ranks, are continually

continually turning their eyes towards it. They are attracting the notice and regard of their neighbours, in their truly noble spirit and conduct.

Let this spirit continue to prevail, let Agriculture be studied by gentlemen of landed property, on philosophic principles; let it be taught to their tenants; and the happy consequence will soon be apparent through this island.

The *difficulty* of instituting Schools for Husbandry is now trifling, since so many Societies have been established, and are supported with so much liberality; especially since the *Society of Arts, Manufactures, and Commerce,* is annually offering such vast sums of money for the encouragement of experiments; and none of the other Societies, I should apprehend, are formed on so small a scale as to preclude the practicability of taking into their hands a few fields, (and a few would be sufficient) and of appointing some person or persons to cultivate them, and instruct the pupils, either according to the idea of Lord MOLESWORTH, which points to the education of poor men's children; or according to the ideas of COWLEY and Sir WILLIAM PETTY, which respect the education of gentlemen's sons as well as others.

At

At present, however, let us attend to the *advantages* accruing from each of the above plans; premising only, that lectures on the theory of husbandry must, *by all means*, be accompanied with a close attention to the practical part of it, in such a manner as may tend to correct the mistakes of speculation, to open and enlarge the mind, and to give a clearer insight into the nature of vegetation, and the very fundamental principles of Agriculture.

Were Schools established in different parts of the kingdom for the education of farmer's sons who might be but in low circumstances, gentlemen would never want sensible and rational improvers of their estates, who would likewise be the most proper persons to instruct parish apprentices and inferior servants. This old experienced VARRO reckoned to be of principal importance. ' The ' bailiffs', says he, ' should be men of some erudi-' tion and some degree of refinement.' But more especially ought a bailiff to be *well skilled* in rural œconomics:* he should not only give orders, but

---

* Qui præsint, esse oportere qui literis sint et aliqua humanitate imbuti,—Præterea potissimum eos præesse oportet, qui periti sint rerum rusticarum: non solum enim debere imperare, sed etiam facere, ut facientem imitentur, et ut animadvertant eum cum causa sibi præesse, qui scientia præstat et usu. Lib. I. cap. 17. apud Authores de Re Rustica. Edit. Jucundi Veronensis, 1529.

alfo work himfelf; that the labourers might imitate him, and be convinced it is with propriety he prefides over them, becaufe he excels them in the practical part, as well as the fcientifick.

Were this the cafe with us, local and eftablifhed cuftoms would be regarded no farther than they are founded in propriety; younger fervants would be accuftomed to a variation in their methods of culture, as *circumftances* varied; new modes would not be defpifed becaufe they *are* new; the effects of experiments would be modeftly expected; the advantages and difadvantages attending them, would be accurately difcerned; and a continual progrefs would be made in the fcience and practice of Agriculture. Were fome fmart boys felected by each Society, and educated on the above plan, they would hereafter convey knowledge wherever they went; and their obfervations would be better attended to by inferior fervants, than if they came from perfons of high rank. In fhort, *they* would effect what even the fuperior knowledge of noblemen and gentlemen could not perform, who have more important objects in view than to cultivate the neglected underftanding of every ruftic labourer they may have occafion to employ. Like fmaller rivulets, branching from the main ftream,
they

they would water and fertilize those lands where a larger river cannot with propriety expand itself.

While under tuition they will learn the expediency of a clean and spirited system of husbandry; as it is supposed that their tutor's fields will be cultivated on these principles. On comparing his crops with those of many others, the truth of HESIOD's maxim would be apparent, that *half may be more than the whole.** For should they think of becoming tenants, they will view an estate with this ruling principle, that one of an hundred pounds per annum, well cultivated, will produce, at the end of the term, more clear profit than another of two hundred a year, treated in a negligent and slovenly manner.

An injudicious course of cropping, imperfect tillage, partial and improper manures, are not always to be attributed to ignorance, but sometimes to the estate being too large for the farmer's capital; he does not command the estate, but the estate him, too frequently to the great injury of both; his hands are bound at his first setting out; and it is much if they regain their freedom, unless eventually through his landlord's distraining him for rent,

---

* Πλίον ἥμισυ παντος. *Opera et dies*, v. 40.

and

and ejecting him from the premises. But what is the farmer to do, if he cannot find a farm in his own neighbourhood suitable to his capital? Shall he remove into another county, an entire stranger, or commence day-labourer, or starve?

The modern practice of throwing several small farms into one, is much to be lamented as a national evil in every view; and calls loudly for the regulation of the legislature.

But to return to our young farmer, transplanted from the nursery, where his mind received its first cultivation, unto the spot where he is supposed to fix his residence.

While under instruction, he was taught to form a pretty good judgment of the qualities, such as the tenacity, dryness, or moisture of different fields, from the herbage they spontaneously produce; he will, therefore, immediately perceive which are most proper to be *first* under tillage, in order that *the estate may not be impoverished*. The want of attention to this circumstance has kept many a man poor all his days, under a notion that the best ground will carry one or two good crops of exhausting corn at first, and so far prove of immediate great gain; not considering that it generally proves a future a future heavy lofs, from the neceffity he will be under of letting it lie fallow, and of applying much expenfive labour in order to extirpate weeds, and much more expenfive manure in order to recover its loft ftrength. Yet ftill, there is a certain vigour in thofe fields, which have been under a judicious courfe of meliorating crops, though but moderately manured, which even a fallow and a complete ftercoration cannot beftow on any foil which hath been once impoverifhed; as may be more eafily perceived by a difcerning eye, than defcribed.

Our farmer hath been taught, that the good ground (on which his chief dependance is for paying his rent) if preferved in good heart, will often mend the bad; but the impoverifhing of one or two of the beft fields will frequently affect the whole eftate in the decreafe of its pafture, in leffening the quantity of manure, and increafing the expence of tillage.

It hath frequently been inculcated on him, That his future fuccefs depends much on his firft courfe of crops; that at firft efpecially, meliorating crops are to be preferred, as far as circumftances will admit, to exhaufting ones;—that the latter, whenever they are fown, fhould be fucceeded by the former;

that

that those manures which are most apt to produce *weeds*, should either be laid on pasture, or ploughed in for such crops as can be best hoed, or have the best tendency to destroy them, viz. beans, pease, turnips, cabbages, &c.—that although some of these crops may require rather more expence, and not return that expence in money quite so soon as some of the exhausting ones, (part of them being appropriated to the fattening of cattle, by which means the best of manure is raised and in the largest quantity) yet, like those bees which travel farthest, and stay out longest, they generally return home most deeply laden;—that the dung-heap be most sedulously regarded as the foundation of his future wealth;—but that no manure should be laid on wet springy lands before they have been drained, unless he chuses to sink the profits of all his other fields.

He hath been taught to venture on some few experiments, on general fixed principles; which, though they might not all of them perfectly answer his expectations, may, nevertheless throw additional light on the subject of Agriculture. In a word, he will become fit company for a gentleman; he will receive and communicate information; and at the same time, on account of that close attention which he finds requisite, in order that he may pay his rent,

he

he will be continually increasing that important knowledge which an uninstructed mind cannot possibly attain.

Such an institution as is here recommended may possibly be of service to those farmers who have no particular connection with our Agricultural Societies; whose fields, however, lying open to the continual view of their neighbours, will be a constant lesson to those who most need instruction, speaking much more intelligibly to *them* than accounts of experiments stated on paper; against which they will be frequently starting that particular kind of doubt, which I have found to be generally expressed in some some such language as this, *It may be so, but I don't know:*—a doubt arising from a cloud inveloping their minds, which the powers of reasoning are very ineffectual to dispel. But they will sometimes learn that lesson from the plants of the field, which they might not chuse to learn from the tongues of their fellow-creatures, because they will not avowedly acknowledge others to be their superiors in this art and science.

The advantages of such an Academy for the education of Gentlemen's sons, will be no less evident with regard to themselves, their posterity, and the nation in general.

On

On this part of our subject, my learned master thus expresses himself:—" According to the best observations, the proper time to infuse that useful part of natural philosophy called Husbandry, is in the earlier stage of life, when there is curiosity and a thirst for knowledge. And if practice here could be joined with theory, enjoying the open air, exercise, and activity, agree well with the turn and cast of young people, not to mention a revolution of perpetual variety which is very engaging at their age.

" It is one point gained, without doubt, to be enabled to read the husbandry works of CATO, VARRO, VIRGIL, and COLUMELLA, with taste and knowledge. It may open a new walk on classical ground; and, in all probability, give young men certain predispositions in favour of Agriculture. Yet still, the whole combined together will produce but slight effects, unless we call in the assistance of facts and experience.

" Something of this kind ought certainly to be done, and the complaint of COLUMELLA, when he says with some degree of warmth, ' Agricolationis ' doctores qui se profiterentur neque discipulos cog- ' novi,'* should, if possible, be removed."

* Harte's Essay I. p. 157.

The

The former part of this quotation evidently intimates, that the improvement of young gentlemen in claſſical learning would not be impeded, but rather promoted, by attending to Agriculture; and the experience of every one who has led a ſtudious life will teſtify that the open air invigorates the mind and prepares it for receiving inſtruction, becauſe it can bear application only to a certain degree, and ſtands in need of being frequently reinvigorated by amuſements and lighter ſtudies.

Time is precious, and might be virtually lengthened by a proper diſpoſal of it. When the mind is fatigued with cloſe application, exerciſe in the open air will renew its ſtrength and activity. Additional to their being taught the value of the different fields over which they may walk with their tutor, from the various plants each field naturally produces, Botany may be attended to as a pleaſing and inſtructive ſcience; neither ſhould planting and gardening by any means be neglected; nor the art of ſurveying and delineating eſtates be conſidered as beneath their notice.*

In bad weather they may be occaſionally amuſed with experiments on various branches of natural

* It is not meant wholly to exclude the ſons of poor men from theſe ſtudies.

philoſophy;—

philosophy;—the effects of the air with regard to vegetation, and the nature of different earths and manures, after the manner of the Doctors HOME, FORDYCE, AINSLIE, PRIESTLEY, &c.

They should also be instructed in the principles of Mechanics, especially that part which relates to Hydraulics, it being of principal utility in draining and other modes of improving estates.

These are circumstances from which many of the capital improvements lately made, in a great measure, originated. They were indeed considered of principal importance by Sir WILLIAM PETTY, ' one ' of the greatest men of that or any other age,'* who recommends them with earnestness, for reasons highly worthy of himself, and which will be mentioned hereafter.

Having gained some knowledge of Agriculture, they will read the works of the ancient agricultural writers with improvement and pleasure; a circumstance which will much expedite the knowledge of the languages. For without excluding other prose authors, may I not venture to assert, that the ancient writers on husbandry are, from the nature of their subject and their classical stile, as proper for

* Biog. Britan. Article Boyle.

young

young perfons, and as fuitable to their difpofitions and capacities, as any they generally read? Indeed I have always been apt to fufpect, that putting the works of Homer, Horace, Virgil, Ovid, or in fact any other *poet*, into the hands of boys, before their minds are properly furnifhed, and their tafte and judgment fufficiently advanced to enter into the fpirit of thofe excellent writers, has been only rendering learning irkfome to them, and proved the means of their bidding a final adieu not only to thofe authors, but to all claffical literature, when they have left their grammar-fchools; not to mention that *profe* writers feem, in themfelves, beft calculated to teach any language by, as well as to convey the moft ufeful information to the minds of youth.

Poetry and painting are fifter arts; they alike receive advantages from rural fcenes: witnefs the fix paftorals of Mr. SMITH, than whom, as a landfcape painter, and as a poet, this age hath not, perhaps, produced a greater.

The following is one inftance, among many others, to prove how favourable an intimate acquaintance with rural images is to poetical defcription:

"The

> "The night was still—the silver moon on high
> "Dappled the mountains from a clouded sky,
> "Silent as fleecy clouds thro' æther sail
> "Before the gentle-breathing summer's gale;
> "So through the misty vale in twilight grey,
> "The sleepy waters gently pass'd away."

Engaging in rural concerns will strengthen the whole human frame, the powers of the mind, as well as the members of the body; will give a manly turn to thought, duly regulated and refined by polite literature. A person thus educated will never want a variety of entertainment in the country to fill up his time in a manner equally innocent, rational, and useful. He will be continually increasing in valuable knowledge, and preserve himself from that dissipation which enervates the mind, renders retirement burthensome, and the more public and momentous concerns of life too arduous to be executed with propriety and decorum. He will enjoy his *otium cum dignitate*, and, at the same time, his private amusements will give a certain dignity and polish to his sentiments, which on all occasions he will be the better enabled to express in public, with a truly British spirit, Roman firmness, and attic elegance. There will appear in his whole manner and address that *simplex munditiis* which is equally removed from empty affected foppishness and

and mere clownish rusticity. He will be fitted for such department in the government of the state as may best suit the natural bent of his genius, whenever his assistance may be thought necessary; and may rank hereafter among those worthies who have acted the same part before him, and whose eulogium may be delivered in the words of the Roman orator: " Ab aratrô arcessebantur qui consules fie-
" rent—Suos enim agros studiose colebant, non
" alienos cupide appetebant, quibus rebus, et agris,
" et urbibus, et nationibus, rempublicam atque hoc
" imperium et populi Romani nomen auxerunt."*

But to return into the more humble walk of cultivation and emolument.

When our young pupil shall come to the possession of his paternal estate, he will immediately perceive what is to be done to the best advantage; he will be able to *direct* his servants, rather than be *imposed* upon by them, which must ever be the case when the master is unacquainted with the business he superintends. This is a matter of high importance. For if in any other profession he should spend his fortune, it is possible he may be the *only* sufferer; but it is not so in Agriculture. Every

* Orat. pro Ligario.

field is, in some respect, public property; and if his crops fail through unskilful management, whatever is lost by the owner is, in some degree, a loss to the community at large.

When I reflect on this, and consider how much the crops are diminished through the mistaken notions and obstinacy of the common farmers, especially when they rent larger estates than they have strength to manage; and when I view the almost immeasureable quantity of improveable land which yet remains waste and next to barren; I cannot but agree to the supposition of Mr. HARTE, that the lands of England may be made to produce one-sixth part more than they do:—a point this of great national importance, amounting to near four millions of money annually!

Whatever advantages may accrue to Gentlemen from committing their estates to the management of such a skilful and well-educated bailiff as hath been above recommended, yet they should not be left *wholly* to him; for experience hath too often shewn, that the integrity of a man's heart does not always keep pace with his understanding.

Indolence, self-interest, pleasure, and other temptations, may cause him to neglect his master's interest

at

at a critical time; the evils of which neglect may not be remedied for years together. Every one who hath attended to works of husbandry must be sensible, that in all their several parts they are only links of one chain; either of which being broken, the whole work is frequently thrown into confusion, particularly with regard to the most proper seasons for the different labours of the field;—a circumstance of no small moment in our varying climate.

This sentiment should be impressed with all possible energy; and it cannot be done in more forcible and comprehensive terms than those of Cato: " Res rustica sic est, si *unam rem* serò feceris *omnia* " *opera* serò facies."

It is likewise to be observed that, although the Gentleman's crops may, in many instances, be larger than those of other men; yet, by trusting too much to his servants, he is often put to needless expence, which the common farmers avoid, and on account of which they object to the propriety of his method; so that hereby the public-spirited gentleman sometimes hurts the cause he intends to serve.

However, the well-educated bailiffs are more likely to do their masters strict justice than the illiterate;

illiterate; those little meannesses which the latter hardly think any thing of, though frequently attended with considerable disadvantages, the former are in general above committing, because they know better.

Indeed I cannot consider the study and profession of Agriculture as any way unbecoming the character of a Clergyman; he may hereafter prove of great service to his country parishioners, as his advice and method of proceeding would be readily attended to by the younger part of his parishioners, and he will have frequent opportunities of conveying just ideas of improving their modes of cultivation.

Thus the knowledge of Agriculture may be diffused in every part of the country, where such a gentleman fixes his residence.

Should his cure be but small, he will have a fair opportunity of preserving himself from that dependance, which might too often lessen the weight and energy which should always accompany his religious instructions.

It was thought proper to reserve Sir WILLIAM PETTY's *Advice for the advancement of Learning*,*   for

* Published in 1648.

for this place; because his plan is in itself highly judicious, and includes the ideas of COWLEY and Lord MOLESWORTH.

Sir WILLIAM proposes,

" That there be instituted literary work-houses, where children may be taught as well to do something towards their living as to read and write.

" That the business of education be seriously studied and practised by the best and ablest persons.

" That all children, above seven years old, may be presented to this kind of education; none being excluded by reason of the poverty and inability of their parents; for hereby it hath come to pass, that many are now holding the plough, who might be made fit to steer the state.*

" That all children, though of the highest rank, be taught some genteel manufacture, in their minority, or turning of curious figures, &c. limning and painting on glass or in oil colours, botanics and gardening, chemistry, &c. &c.

* Cincinnatus was called from the plough, in order to steer the state as Dictator; and returned to it again after he had delivered Rome from her danger.

" And

"And all for these reasons:—They shall be less subject to be imposed upon by artificers: they will become more industrious in general; they will certainly bring to pass most excellent works, being, as gentlemen, ambitious to excel ordinary workmen. They being able to make experiments themselves, may do it with less charge and more care than others will do it for them. It may engage them to be Mæcenas's and patrons of arts. It will keep them from worse occasions of spending their time and estates. As it will be a great ornament in prosperity, so it will be a great refuge and stay in adversity and common calamity."

After these observations, need any thing be added to shew the advantages of such an education, except attempting to obviate an objection which may possibly arise with regard to the difficulty of procuring proper tutors?

This, however, seems to be a difficulty, which, in this enlightened age, may be soon surmounted. I imagine there are many persons in the kingdom well skilled in scientific and practical knowledge, who would, were they encouraged, readily step forward, and reduce Agriculture (both in theory and practice, with all its connections and dependencies

on

on botany, chemiftry, and other branches of natural philofophy) into a fyftem of education as regular, plain, and introductory to right conduct, as in any other art or profeffion in life. Let it but have a beginning, and inftructors would, no doubt, foon abound.

The author will not be wanting in any thing which lies in his power, however fmall, to promote fo defirable an end.

And with this declaration he fubmits the foregoing obfervations to the judgment of the wife, the candid, and benevolent.

ADVERTISEMENT.

IN order to render the study of Agriculture more general, especially among the rising generation, the Author proposes, should it be thought eligible, to publish, for the use of schools, an Abridgement of the writings of CATO, VARRO, and COLUMELLA; by selecting such passages as seem more especially adapted to the husbandry of these kingdoms, and to be of public utility.

This advertisement owes its existence to Mr. AIKIN's edition of *Selecta Quædam ex Plinii Hist. Nat.** which cannot but be considered as a very valuable addition to our small store of Classick Authors, proper for the use of schools.

It is thought that a judicious selection from the writings of the three Authors above-mentioned, if well *translated*, would be of service.

* Select passages from Pliny's Natural History.

A LETTER

TO

MONSIEUR HIRZEL,

FROM

DOCTOR TISSOT;

IN ANSWER TO

MONSIEUR LINGUET's TREATISE

ON

BREAD-CORN AND BREAD.

PRESENTED TO THE SOCIETY IN FRENCH

BY SIR JOHN PRINGLE, BART. P. R. S.

AND TRANSLATED BY A MEMBER.

THIRD EDITION.

1792.

TRANSLATION

OF A

Letter from Dr. Tiſſot to Monſ. Hirzel.

HAVING frequently received much information from the works of Monſ. LINGUET, I always read them with pleaſure and with full expectation of further improvement; but, however well grounded ſuch an opinion of any author may be, it ſhould never go ſo far as to prevent a ſtrict and impartial examination of facts. The examination which I have made of his Treatiſe upon Bread-Corn and Bread,* does not permit me to adopt his opinions on two ſuch intereſting ſubjects to mankind. I even think it might be of dangerous conſequence, ſhould they become general; and when an author of ſo much genius, learning, and eloquence, undertakes to eſtabliſh an *opinion*, however abſurd, it may probably bias the judgment of ſome part of his readers, and be a means of per-

---

* Annales Politique, Civiles, et Literaire, tom. v. p. 429.

ſuading

fuading them to adopt the fame fentiments; I therefore thought it might be ufeful to publifh the reflections which I made in reading this feducing Treatife.

I fubmit them to you, Sir, as to one of the moft competent judges, engaged, both by ftation and natural abilities, in every thing that tends to the enriching your country, and the welfare of your fellow-citizens; profoundly verfed in all the branches of Œconomics, Agriculture, and Phyfic, you will be equally capable of difcuffing the objections of M. LINGUET againft the ufe of Bread, and my obfervations upon them; your decifion will certainly have very great weight in the fcientific world.

Monfieur LINGUET affirms, that the culture of bread-corn is prejudicial, and that bread is an unwholefome food. The latter of thefe principles only can be properly confidered in a medicinal view. However, I muft be permitted to examine the firft alfo; fince it would be of very little importance to defend the ufe of bread, if the culture of the grain which produces it be prejudicial.

It is a certain fact that, in fome countries, one arpent of land, fown with corn, yields lefs than
the

the same quantity planted with vines, or of good meadow-land; and according to the manner of reckoning, a district which had one thousand arpents* of arable land, would receive less profit from its produce than that which had one thousand arpents of vines, or one thousand arpents of meadow-land; however, this is not owing to the corn, but to the soil; for they sow it with corn, because it is not good enough for meadow-land or vineyards; and if the profits of an arpent of arable be less than that of meadow-land, it is because the soil of the one is not so rich as that of the other. If corn were sown on a soil naturally good, without the help of manures, I am persuaded that more advantage would always accrue from arable lands than from meadow. The same comparison cannot be made with respect to vines, because they must have a particular situation. But corn is more easily cultivated than grass; for although they are two plants of the same species, the former will

---

* I am obliged to use the word *arpent*, as our English acre does not answer to it; and I know no other word in English applicable. The common arpent in Switzerland is called a *pose*, and measures 40,000 square feet; the *arpent* of Paris, 100 perches, reckoning 18 feet to the perch, is 32,400 square feet; but as the foot of Berne is less than that of Paris, in the proportion of 1500 to 1440, the arpent of Paris contains 36,735 feet of Berne; and as the difference is but 2265 feet, one may be taken for the other, without any error of consequence.— *N. B.* The English acre contains only 40 perches.

thrive in lands where the latter will not, or, at least, it grows so weak and thin, as to be easily over-run with weeds, or dried up by the heat of the sun; it has therefore been found necessary, in districts where the land is not good, (which is most commonly the greater part) to leave the best for hay, and to put the corn into that which is but indifferent, or even in the worst of all; and though they cannot expect very great crops, yet they reap something.

If there are some districts of very poor lands, almost entirely sown with corn, they are not poor, because they produce only corn, but because they are not fit to produce any thing else. Their soil is so bad, that they can grow but very little fodder, consequently they maintain only such cattle as are absolutely necessary for labour, and those are ill fed, and frequently perish. They have but little manure, and their crops are small; for large crops of all sorts can only be expected from lands naturally rich, or strongly manured. Thus the poverty of the inhabitants is only owing to their possessing an ungrateful soil.

What proves evidently it is the natural soil that is in fault, and not the corn which impoverishes it, is, that where there is meadow and arable land, the

price

price of the meadow land is much more considerable than that of the arable. In moſt parts of this country,* the proportion is nearly ten to one; and there are even ſome arpents of meadow, for one of which they would give thirty of field lands, and ſome of vines for which an hundred of arable land would be given.

Thoſe diſtricts, where the ſoil will produce nothing but corn, are poor; but in thoſe which furniſh fodder, and alſo fine crops of grain, the inhabitants are wealthy and happy, unleſs they are oppreſſed by taxes.

There are many inſtances of this kind in this country, which Monſ. LINGUET has not given himſelf time enough to conſider with proper attention; and ſurely, it is moſt probably ſo in other countries. Flanders, Brabant, ſome parts of Germany and Poland, Milan, and England, which furniſh great quantities of grain, are countries abounding alſo with all the neceſſaries of life, enriched by the money which the exports of their corn bring in return. If there are many poor in them, it is not their raiſing corn that occaſions it; but the unequal diſtribution of it. Whatever commodity a country

* Switzerland.

produces,

produces, if it is not enjoyed as private property, but is reaped for others, the inhabitants still continue poor.

In some provinces there are lands of very considerable extent sown with corn, which belong to the church, or perhaps to some nobleman. The peasant may be poor in the midst of this opulence; but it is not because there is corn, but because it does not belong to him. If there are countries where they reap plentiful harvests, and where, nevertheless, the owners themselves are poor, this poverty is not owing to that plenty, but to some other cause; frequently, perhaps, their situation is unfavourable for vending their grain, and then, undoubtedly, it would be better to sow less of it; perhaps, indeed, (almost universally) too much land is appropriated to the culture of grain. If less were cultivated, and the husbandman would be more attentive to the cultivation, better crops might be produced at less expence: thus the advantage would be much more considerable:—but I shall speak again of this hereafter. However, this proves nothing against the cultivation of bread-corn, since, if it be cultivated with care, the produce will be very considerable. If farmers in general sow more than double what is necessary; if they sow it only

in

in very poor lands; if thofe lands are badly prepared, and they do not allow the neceffary quantity of manure; it will be with wheat as with all other crops; it will not grow, becaufe it has not been properly cultivated.

You know, Sir, that the experiments of Mr. TULL, DUHAMEL, MOUGRES, and many others, have demonftrated the advantage of fowing much lefs feed than is ufually fown. Perhaps you recollect, that this method, any more than the ufe of the feed-bag, is not a new difcovery, but has been proved by experience more than an hundred years.

In the Philofophical Tranfactions for 1670, No. 60, we find a very full and particular memorial of Mr. EVELYN's; in which, after the Spanifh Memoir of M. le CHEVALIER LUCATELLO, he gives the defcription of a feed-bag ufed in Spain, called a *fembrador*, which the inventor, after having fully eftablifhed its great utility, by repeated trials, in the prefence of the Emperor, took into Spain, where the government ordered feveral new trials to be made, which were alfo attended with great fuccefs. By this means, two-thirds lefs is fown, and they reap more. The care required in the conftruction of the plough, to which the feed-bag is adapted,

adapted, and the work it requires, are explained very clearly; and it is very probable, that it was from thence that Mr. TULL has drawn his difcoveries. One finds alfo in the fame publication, that about the year 1665, the Royal Society appointed a committee, who employed themfelves in enquiring into every thing relating to the hiftory and progrefs of Agriculture in thefe kingdoms: This committee publifhed queftions the moft interefting, and the beft calculated to anfwer their defigns in enquiring into all the different branches of Agriculture, in order that from a knowledge of the true ftate of it, and from the obfervations of perfons fkilled in œconomics, whom they requefted to communicate their fentiments, they might fully eftablifh that part of it which feemed to them of the greateft importance. Thefe queftions contained almoft every thing that has been propofed fince that time: and it appears that this committee were employed without being much known, on the fame objects which have engaged the attention of all Europe for twenty-five years paft, with fo much enthufiafm and oftentation.

But to return to my fubject. Suppofing the common increafe of wheat to be fix and a half, as it is generally fown at prefent, this would be thirteen

to

to one, if only half that quantity were sown, and this would be a very fine produce.

M. LINGUET has, I think, gone too far in supposing that the culture of wheat requires more time than it really does. One arpent of wheat requires no more than four days' labour in the year for sowing, two for reaping, and the same for manuring; let us then reckon six for threshing, and two for grinding it; and this, in the whole, makes sixteen days, which is all that it requires. Let us then suppose a family, consisting of six persons, (three men and three women) three arpents would supply them with more corn than would be sufficient for their sustenance, and would require only forty-eight days' work; and even of these forty-eight days, it would be only those of harvest that would employ the women; those of cleaning, weeding, or halling, if necessary, would employ them but two, and they would have nothing to do with the sowing or manuring. The grinding and baking take up but little of the men's attention, neither would they be always employed in the other three parts of the work; therefore all the remainder of their time may be employed in other occupations. I am well aware, that if more land be cultivated, it will necessarily require more time, although the time neceſſary for the culture of arable lands does not increaſe in proportion to their extent; but in that caſe, the extraordinary time employed is making a trade of the produce, and not that which is barely requiſite to acquire a neceſſary ſubſiſtence; and this may be increaſed in any degree, even till their whole time would not be ſufficient.

Water-meadows, which alone may be deemed truly fertile, require daily care to water them at leaſt ſix months in the year; and the harveſt alſo requires much care. The culture of vines requires much more attention and time; and it is therefore ſuppoſed, that if a Vigneron can take care of a certain number of arpents of vines, the farmer can, with the ſame time and trouble, attend to a farm eight or ten times as large.

I know very well, that the one requires cattle, and the other does not; but theſe cattle, far from being expenſive, will, if properly managed, increaſe the gain of the farmer; therefore, they muſt not be looked upon as an expence.

Corn is ſubject to many accidents, but vines are ſubject to many more; and thoſe which the vine ſuffers, ſometimes ſpoil the vintage for ſeveral years;

years; thofe which happen to arable land only fpoil the harveft for the enfuing feafon: And as the expence of cultivating vines, for which only manual labour can be employed, is much more confiderable; therefore the Vigneron, who engages more largely than the farmer, will confequently be a much greater lofer, if unfuccefsful.

Hay is alfo fubject to frequent and very difagreeable accidents: the fecuring it is fometimes very difficult; and when it is badly made, it becomes very hurtful to cattle. A fingle fact will be fufficient to prove the cafualties hay is fubject to, which is, that it varies in price as much as grain. Accidents of hay-mows taking fire are but too frequent, and this is not to be feared in corn-mows.

The prefervation of vines is not attended with lefs difficulty than that of grain, and the accidents they are liable to, being more fudden, cannot be fo eafily prevented. When grain has been well taken care of in the harveft, it is very feldom any accidents happen to it, except the mows are made very large; and even then, every one knows there are certain means of prevention. It may be faid with truth, that if wheat does not keep, it is the fault of the farmer: Nature has formed it for long prefervation;

preservation; and this is so well confirmed by repeated observations, that it is no longer to be doubted, that wheat is the most durable of all eatable grain.

The care it requires in grinding is common to all other sorts of grain; and the accidents which they are liable to are as many as those of wheat: their produce is not so certain; and, finally, the straw of wheat is of more general use and importance than that of any other grain.

Maize and Millet impoverish land very considerably; and in those countries where maize flourishes the best, the owners, when they lease out the farms, particularly specify, that only such a portion of the arable lands shall be sown with maize: and in Piedmont, one of the districts where they cultivate the most, this proportion is not to exceed the forty-eighth part.

With regard to Rice, every one knows, that it will not grow without being under water; and that four months out of the six that it is in the ground, the land must be kept flooded; consequently the countries where rice is cultivated are very unhealthy; and in the county of Verceil, a man of forty years of age is old and decrepid,

The

The culture of Rice* is, therefore, detrimental, and far from being proper to be encouraged in any state; they ought rather to guard against its being introduced; for even though it might be very advantageous in a lucrative view, yet it is certainly of more confequence to preferve the health and lives of the inhabitants, than to feek the means of enriching them. The sovereign council of Roussillon prohibited it about forty years since, becaufe they thought that the exhalations of the lands fown with it had occasioned an epidemic distemper.† Another argument against the culture of rice is, that it requires a confiderable degree of heat as well as moisture, and therefore will not grow beyond the latitudes of 46 or 47 degrees, and perhaps even not so far.—Mr. Rye, a very accurate obferver, has affirmed, that it diminishes by transplanting; therefore, if it were advifeable in countries where there is great plenty, it would not be fo where it is already thin.

* The culture of Mountain Rice would, no doubt, be very advantageous; but befides the great uncertainty of its growing in Europe, the procuring it is attended with much difficulty, fince the zeal and ardour of Monf. POIVRE, who firft made it known to us, hath not been able to furmount it.

† Thofe inhabitants of the mountains, who come down into the vallies of Piedmont in the autumn, to the Rice harveft, are moftly feized with the fever.

It appears then from what has been said, and which may also be relied on from a great number of treatises on the subject, that Wheat is not a commodity that is impoverishing in itself; for, in an equality of soil, it is as advantageous as any other; and that this grain will grow indifferently, at least in lands and situations which are unfavourable to other plants. One may also add, that this grain is adapted to most climates; and that, if there are districts almost entirely sown with wheat, and yet poor, it is the fault of the soil, or some other circumstances, and not of this useful grain.

The comparison between the population of some provinces, cultivated with corn, which are thinly inhabited, and others of vineyards and pasture-lands, which are more populous, simply proves this only, That one soil is more rich than the other, and that a fertile soil will maintain most inhabitants.

No person is more capable of assigning the cause of the subjection of the Roman empire to the Northern powers, than M. LINGUET; but he cannot surely be serious, when he says, that they were enabled to conquer them, because those Northern countries produced no corn, and that population decreased since the introduction of grain.

I shall

I shall make three observations on this passage.

First; The armies of Gustavus Adolphus, Charles the XIIth, and the King of Prussia, whose food was bread, would be as formidable against the Italians of these times, (who eat less than was eaten in the days of Scipio) as their ancestors were fourteen hundred years since against the Romans: And as M. Linguet speaks of conquests, he must know certainly that those Greeks who subsisted on bread, those Romans who ate nothing but bread and vegetables in pottage, subdued all the known world, amongst whom were many nations who ate less bread than themselves. The ration or allowance of bread for a Roman soldier was much more than what soldiers have at present; and they were also much stronger: The allowance to a Roman soldier was 64 pounds of wheat per month, which he was strictly forbidden either to sell or exchange: They had very seldom any cheese, bacon, or pulse; wheat was then almost their only food, and the proportion was double what is allowed the soldiers in our days: they ate it in bread, in flour-milk, and in thin cakes, and they were not subject to epidemic and putrid disorders, as is but too much the case with our armies at present. Bread-corn then did not diminish their strength, as one may

judge

judge from the weight of the accoutrements, which they carried, neither did it make them less brave, or in any degree unhealthy; nay, it is very probable, that the most certain method of preventing epidemic diseases in the army, where it is so difficult to procure good animal food, would be to reduce them to the simple diet of the Roman soldiery.

Secondly. It is very doubtful whether those countries were more populous, than they are at this time; it is even very probable, that they were less so.

Finally. The people of these Northern countries were not without wheat: it was the basis of their food and their drink: without quoting other authors who attest it, suffice it to say, that TACITUS affirms it in his *De Morib. Germ.* cap. 23, 25, 26.*

Monf. LINGUET's second remark is, that of nine hundred millions of men, there are scarcely fifty millions that use corn for their food; but in this he

is

---

* Portui humor ex hordeo aut frumento, cap. 23. Frumenti modum dominus injungit, cap. 25. Agri per vices occupantur, cap. 26. Non contendunt ut pomaria conferant et prata feparent et hortos rigent sala terræ seges imperatur.——Corn then was the only object of their culture: and milk boiled with flour, wild apples, fresh game, and curdled milk, appear to have been their principal nourishment or food.

is certainly guilty of a very great mistake, either wilfully, or through inadvertence; for, although there may be some small districts in Europe where

NOTE BY ANOTHER MEMBER.

* TACITUS's words are, "Cibi simplices agrestia poma, recens fera, et lac concretum." From this M. TISSOT concludes, that the basis of their nourishment was corn, which is not once mentioned. It is true he has added, *bouillies de cetteferine*; but this is a mere interpolation of a French dish, and not to be found in TACITUS. It is probable they made use of corn more for the purpose of brewing drink, than for solid food. What TACITUS says negatively, M. TISSOT has interpreted positively; he says, indeed, that all they raised from the ground were crops of corn, and that they neglected the culture of fruit-trees, and improving meadows. But their culture of corn must have been very little, when they never cultivated the same tract for two years together, and still there was a superfluity of land. "Arva per "annos mutant, et superest ager." TACITUS also says, that the whole wealth of the people consisted in their cattle; and that all their fines and mulcts were paid in cattle. "Sed et levioribus delictis pro "modo pœnarum, equorum pecorumque numero convicti mulctantur." Cap. xii.—" Luitur enim etiam, homicidium certo armentorum, ac pecorum numero." Cap. xxi.

Besides, it is highly improbable that a people, who lived upon corn, should have no word in their language to express the autumn or harvest time of the year, as TACITUS expressly says, "Unde annum "quoque ipsum non in totidem digerunt species, hiems, et ver, et "æstas, intellectum ac vocabula habent, autumni perinde nomen ac "bona ignorantur."

But there is another proof more directly to the point, which is, that JULIUS CÆSAR, in his VIth book, giving an account of the Germans, says expressly, that they did not mind Agriculture, but lived on milk, cheese, and flesh meat; and that Agriculture was purposely discouraged among them, lest it should lead to effeminacy. Add to this, that a diet of flesh is much more suitable to a cold climate, than one of vegetables.

Had M. TISSOT read these passages with accuracy and candour, he never could have made so absurd an assertion.

rice,

rice, maize, buck-wheat, and *chesnuts, are the principal food of the inhabitants, yet one may aver, that there is not the smallest province, if we except a part of Lapland, where corn is not the basis of their nourishment. Europe certainly contains not only fifty, but one hundred and twenty millions; and it is not in Europe only that corn is the principal food of the inhabitants.

Monſ. LINGUET makes an exception to this valuable grain, by saying it will not ripen but in the latitudes between 25 and 60 degrees; but it is precisely in these latitudes that population flourishes the most; it is there that mankind increase the fastest; the number of the inhabitants of the Torrid and Frigid Zones does not bear any proportion to those of the Temperate Zones; and the suitableness of these climates being more favourable both to men and corn, is a strong indication, that the one is destined for the other; besides, there are many places in the Torrid Zone where bread-corn is cultivated; it may therefore be fairly presumed, that it would grow in many others; and I would boldly ask Monſ. LINGUET, what other eatable

* The use of chesnuts, as food, diminishes daily; and it is a complaint of the œconomists of France, that the chesnut-trees are much destroyed in those provinces where they used to be in the greatest plenty, in order to plant mulberry-trees in their place.

grain

grain is accommodated to such a number of climates? and, above all, what grain is so generally known and made use of amongst so great a variety of nations? It is found in all Europe, in Egypt, and throughout Barbary, (that is to say, in all the most populous parts of Africa) in Mexico, in the most flourishing provinces of Peru, and in those of North America; it is the principal food of the English colonies; and for some years past, America has been enabled to export quantities of flour to Europe. Wheat is also the food in many provinces of Persia, Mogul, and Indostan; and it is found in all China, where there are three provinces that have no rice. Thus, you see, it is cultivated by all civilized and industrious nations.

That wheat is not used among savages, or the Arabs, is no argument against it; for even some of these have cultivated it in all ages, as TACITUS affirms, in those countries where the soil will only admit of particular cultures; and, especially in that sort of land which is only favourable to the manioc, the inhabitants are obliged to seek for means to take away the poisonous quality of this plant, in order to make it their food; but notwithstanding several persons are poisoned by it every year: wheat therefore is, no doubt, the general food of all civilized

lized nations; and there is the strongest presumption, that it is also the most advantageous to cultivate; but this would be a most cruel mistake, if it were so unwholesome a food, as Monf. LINGUET affirms.

I am not an enthusiastic admirer of bread: I have even said that bad bread only, or the wrong use of it, may be hurtful. I also added, that in some languishing disorders, even the best bread ought to be used with moderation; and there is no other food that may not be liable to the same objections. But it is nevertheless true, that of all foods, wheaten-bread, well made, is the most wholesome; and I am too much attached to the preservation, the health and the happiness of the people, to be suspected of having given any advice on these three important subjects, without due consideration.

You will see in the sixth volume of my Dissertation on Nervous Complaints, all that I have written to dissuade them from the use of l'Ergot,* a grain certainly poisonous, which some able physicians (deceived by superficial observations) esteemed a wholesome food, and which might be used without risk.

* Spur-corn.

I write

I write now to preserve them against that bad opinion of wheat, which an ingenious and eloquent philosopher (who, from some few particular objections, has drawn a false general conclusion) might incline them to adopt.

It appears to me as necessary that man should have good bread, as it is dangerous to have bad: And it is very singular that there should be, at the same time, well-meaning and learned men, who forbid the first as a poison, while others recommend the latter. Vegetables have always been necessary for man.

No nation has ever been discovered who lived wholly on animal food. All have made use of bread, or some equivalent for it; that is, some vegetable farinous substance, which prevents the satiety of all animal food, and the corruption which would be the necessary consequence: even the bark of fir-trees has been used for this purpose; but I do not scruple to affirm, that of all these vegetable substances, bread made of wheat is certainly the best. Of all the objections which M. LINGUET makes against it, there is not one well founded. It is very certain, that if wheat be simply pounded, as maize, buck-wheat, or millet, in the kneading
it

it and baking it, much better cakes are made than from any other flour. I have already mentioned that the Roman foldiers ate it in all thefe forms. None of thefe grains can be eaten green. Wheat is undoubtedly the leaft difagreeable and the beft; but were it to be ufed without dreffing, it might be attended with danger, and Monf. Van Swieten has feen the bad effects of it: but all the other grains would be more hurtful, even were it poffible to fubfift on them. The bran, which is fo much objected to, is only the outfide or rind: all other grains have fomething of the fame kind, and that of wheat is the only one which is worth preferving.

It may have the fame objections with other grains, of being a vifcous food, (if not made into bread) difficult to digeft, except by the moft robuft conftitutions, and likely to caufe obftructions even to the moft healthy, if they lead a fedentary life; but it has lefs of this quality than any of the other vegetable fubftances which are efteemed wholefome food; and it is much eafier taken off by a gentle fermentation, to which it is peculiarly adapted; and far from injuring the dough, it only takes away the vifcous quality, by clearing it of that fixed air which combines its parts together, rendering it more light and much eafier of digef-
tion,

tion, and consequently more wholesome. This fermentation makes it much better, instead of spoiling it; and it is not more reasonable to suppose, that bread should be spoiled by this method of fermentation, than wine is spoiled, because it is produced from the same kind of fermentation.

What food is so proper, so refreshing, as bread? Nothing cloys so little. If the fermentation be too great, the bread may be a little acid; this is a defect; but even this defect does not make it less wholesome for many constitutions: and M. LINGUET is mistaken, in admitting, that of all digestible substances, there is none more hurtful, ' more hard ' of digestion, or more heavy in the stomach;' he also adds, that ' it occasions the blood to be thick ' and corrupt; and one of the most celebrated ' aphorisms in physic is, that the indigestion pro- ' duced by it is certain death.'—These assertions have been too carelessly advanced on the testimony of some persons who have examined the effects of bread very superficially. Of the flour or bread which is produced from grain, there is none of which one may eat a greater quantity, that digests more easily, or which corrupts the blood so little. Thick blood is the effect of too strong an action of the vessels, or too quick a circulation; bread does not

quicken

quicken it too much, nor does it cauſe that ſlow circulation, like other farinous ſubſtances. A light decoction of bread is a very wholeſome nouriſhment and beverage in agues, putrid fevers, and in the cholera morbus.

The aphoriſm which M. LINGUET quotes above, is an error of the leaſt enlightened times. I dare affirm, that no perſon ever died of indigeſtion from bread; and in oppoſition to this quotation, I ſhall bring an authority more reſpectable than the ſchool of Salernum. You know, ſir, that HIPPOCRATES remarks, that in a ſcarce ſeaſon of wheat, when they were obliged to ſubſiſt on other vegetables, many paralytic complaints appeared, the natural conſequence of bad food and relaxed fibres.

We may further add, that if the ſuperiority of wheat, above all other edible grain, had not been demonſtrated by ſo many indubitable facts, it would be proved by the ſeveral accurate trials of M. M. BECCARI, KERSEL, MEYER, ROELL, and MACKER, upon flour; that the glutinous or animalized ſubſtance, abſolutely unknown till within theſe forty years, which is the fourth part of wheat flour, and of which the flours of other grain have ſcarcely any, ſeems deſigned to render this flour

more

more easy of digestion, and more nutritive, since it loses less in being reduced to a proper state for food, and is more stimulant. I say, this glutinous quality ascertains the superiority of wheat above all other grain; and it seems to me, after many observations, and on a comparison between the inhabitants of those countries which have no wheat, and those where it is the principal food, one may venture to affirm, that it is the nourishment, of all others, most favourable to the mental faculties.

Those whose food is maize, potatoes, or even millet, may grow to a large size, or be tall; but I much doubt, Sir, if any one, who lived wholly on them, could ever write the Political Annals of the Sixteenth Century, the pleadings of Monf. Le Duc D'Aiguillon, and the Defence of Monf. Le Comte De Morangier, &c.

If the inhabitants of Europe are, in all respects, superior to other parts of the world, it is owing, perhaps, to their great use of wheat. If many nations eat less of it than others, and yet appear equal in all respects, it is that the nature of their other food and drink requires less: those who drink beer make use of soaked bread; or, indeed, it may be deemed a kind of bread dissolved, which has the double effect

effect of bread, both as to nourishment, and as an antiseptic, preventing the putrid effects of other food. There are others who eat it under a variety of forms, and, if they eat less bread, may consume an equal quantity of flour; and again, there are others who do not eat enough, and that is perhaps the occasion of those diseases which carry off such numbers. Those districts which consist chiefly of dairy farms have less need of bread than others; and for the same reason, the inhabitants of mountainous countries should eat more.

But Monf. LINGUET should take notice, that it is because milk is of much the same nature as bread, that it yields a nourishment partly vegetable and partly animalized, and that it also contains a substance similar to that valuable glutinous quality of wheat.

And M. MACKER, whose decisions are a law in chemistry, has discovered, that the curd of milk, joined to those particles which contain only the starch, would be one of the best means to render it capable of making good bread.

What still proves farther the superiority of wheat above all other farinaceous grain, it is necessary to
make

make the ſtrongeſt beer; and nothing can be found as a ſubſtitute for it. Is not this union of an animalized and vegetable ſubſtance, in the ſame grain, a certain indication that it is deſigned as the principal food of a being, who, like man, is deſtined to ſubſiſt on animal and vegetable ſubſtances?

If there are men very lean, withered, and decrepid, in countries where they ſubſiſt on bread, this is not, ſir, becauſe they eat only bread, ſince it is known that the Roman legions lived upon it, and were very healthy; but it is becauſe they do not eat enough of it, or eat that which is bad; either it does not afford nouriſhment ſufficient, or the quality is bad; and they are alſo worn out with hard labour.

If there is a ſmall diſtrict in the Pyrenean mountains, where every houſe contains a patriarch, and every garden receives a happy man who ſubſiſts on maize, which does not grow on many other mountains, it is becauſe they are entirely ignorant of our manners, and are wiſe enough not to have any ambition of knowing them; but this way of thinking would render them equally happy with flour-milk as with their Turkiſh corn. In theſe countries the clearneſs of the air prevents the in-

conveniences

conveniences which are occasioned by this kind of of food in other places.

It is happy for the inhabitants of those countries where corn will not grow, that they are able to subsist without bread, and where nothing is produced to give in exchange for those things they want; and it is also a great happiness, that in countries which produce nothing, but where the industry of the people brings in a great deal of money, they can, with that money, import from foreign parts, and sometimes from very distant places, a species of provision which will bear very long voyages, and may be preserved many years unhurt; but most other grains can be kept only a little time, and will not bear exportation so well.

In 1713, wheat sold here at six livres the French quarter, which came to at least twenty-eight livres the quintal; and the quantity that each person was allowed to purchase, was regulated. About eight years ago it was nearly that price, and a like scarcity may be again experienced.

There would have been a famine in Switzerland, if they had not imported grain, not only from Piedmont and Milan, but also from Sicily and Barbary.

Maize,

Maize, which is not reaped till October, is not eatable immediately, unless very nicely dried, (and this drying requires much more care than wheat;) it also spoils very soon; and however well it may be got in, even if it does not appear to be altered, it acquires from the month of June, a considerable degree of acidity, which renders it less agreeable, and less wholesome; besides, the property which it has, of so quickly fattening animals fed with it, proves that it is not so wholesome a food as wheat, which, without fattening them so soon, gives them a firmness and flavour; it appears also by this, that it does not give them a firmness of fibres; and may not this be one of the causes, so well established, that many people in America, who live only on maize, are so inferior in physics and morality to the European nations?

Monf. LINGUET has not had an opportunity of tasting preparations of buck-wheat or millet; if he had tasted them, he would not have advised any person to substitute them in the place of those prepared from wheat: And even if these plants could furnish so agreeable and so wholesome a food, yet they would be liable to many real objections. All the millets impoverish land to such a degree, that if they are often sown in the same land, it will produce nothing elfe for a long time; it is for this reafon, that the culture, very flightly recommended fome years ago, is now abfolutely decried. Buckwheat, the produce of which is fometimes very confiderable, when it grows, has not the fame inconvenience; but it is the moft tender of all plants, and the moft uncertain; the flighteft intemperance of the feafon reduces its produce to nothing; and one may venture to affirm, that thofe countries in Europe, which depend on buck-wheat for their fubfiftence, run a rifk of being frequently threatened with a famine.*

Before I faw M. LINGUET's Treatife, I did not know that there were no poor people in Ireland and Scotland; but I know very well, that if the police does not hinder it, there is a great number in all fertile countries, becaufe the indigent, from the poorer diftricts, will go thither.

I do not believe, and permit me to tell you fo, Sir, that one fack† of wheat takes more from the land

---

* But little is fown in this country, where it is feldom ufed but for fattening poultry; and they fow only the buck-wheat of Britanny. As the months of July, Auguft, and September, when it is on the ground, are often very dry, perhaps it would be better to fow the large buck-wheat, which grows better in dry feafons.

† What is called in this country a fack of wheat, ought to weigh two hundred pounds: And an ingenious phyfician at Lyons, one of my

land than is sufficient to bring up and subsist a poor person; but I know that a sack of wheat would be sufficient for him to live upon at least four months; and I have seen, that where there is one ear of corn to be gleaned, ten poor people who are in want of it, will go from afar to gather it. And if we admit what Monf. LINGUET advances, that there are poor persons who are shamefully obliged to beg their bread on the very furrows which produce it in plenty; in comparing this proposition with the little quantity of corn which is sufficient to subsist a person, it is impossible to suppose, that he is in danger of being starved, because his country produces a plenty of corn.

my friends, has made the following experiments with great precision: 500lb. of ground wheat, without separating the bran, yielded 497lb. of flour, which produced 448lb. of paste, and 430lb. of good bread. A sack which weighs 200lb. will yield 286lb. of bread; and if 20lb. be allowed for the bran being taken away, there will remain 266lb. of very good bread, of which 2lb. per day will certainly be a very sufficient quantity for one person, who, with these 266lb. of bread, may subsist 133 days, or at least four months: Allowing a tenth part for the expence of grinding and baking, it appears that, according to this calculation, a man in an uninhabited island who possessed but three-fourths of an arpent of land, of which he could easily dig up half an arpent to sow wheat, and who could, in the remaining part, cultivate some sorts of pulse, (of which I shall treat hereafter) above all, cabbages, would reap above 800lb. of wheat; he would then have 600lb. to subsist him; and although he should have only 200lb. to sell, yet this would furnish him with what salt and butter he would have occasion for; and the straw would procure sufficient manure; and I much doubt, if this spot would, employed in any other manner, afford him so certain or so wholesome a subsistence.

I hope

I hope what I have said will undeceive Monf. LINGUET, and alter the wrong opinions which some misinformed physicians have given him concerning the bad qualities of bread, which is certainly the most wholesome of all foods; and that in re-examining all the circumstances of those countries which produce it, he will find that these are rich or poor in proportion to the goodness of the soil, and that a plenty of wholesome food can never be the cause of poverty.

If monopoly,* bad regulation, or bad management in the corn trade, bad cultivation, or fraudulent practices, have caused Monf. LINGUET to be disgusted, it is not the fault of the corn itself, in which trade fewer would be employed, and on which still fewer speculations would be made, if it were of less value, or not of real importance. A value for which, could any thing else be substituted by the individual that cultivates only for his own support, it never could be by the minister who has

---

* Monopolies will be carried on in every country, where covetous wretches are found, who are not afraid to attempt it. I have read in the public prints, that a monopoly of hay in one part of the Ecclesiastical Territories, had reduced them to great difficulties in providing for their horses. A monopoly of potatoes, maize, or buck-wheat, might be much easier made than of hay, and money will always induce the greatest number of peasants to sell that in the morning which ought to subsist them at noon.

fleets

fleets and armies to provide for, which could not otherwise be supplied, and also magazines to furnish in case of a scarcity, or unfruitful years; but magazines cannot be formed of provisions which take up a great deal of room, and which must be often changed.

It would be very dangerous to trust to any of those grains for a subsistence, which are subject to more accidents than wheat, and of which even the harvest may totally fail, and whereof one cannot lay up a store to serve in cases of necessity; this would be exposing us to very frequent famines; and certainly that is a very convenient provision, of which the great plenty of some years will more than compensate for those of scarcity.

Besides, bread has the great advantage, when well made and baked, of keeping a long time, of bearing exportation, of being always ready without any fresh baking; this is also a very valuable property, and perhaps is a property peculiar to wheaten bread, since other compositions with paste, unfermented, will not keep near so long.

I should still have further remarks to make on other subjects of this Treatise, but I do not like to write

write so long in contradiction to this author. I shall therefore conclude, with pleasure, by speaking of a subject on which our sentiments are nearly the same, and that is, POTATOES. I am persuaded, and I have mentioned it in a work nearly ready for publication, that there are few kinds of food so wholesome: and there are none of the farinaceous kind unfermented, of which one may eat so much. I think them far preferable to maize, buck-wheat, millet, or even rice; and one may eat almost as much of potatoes as of bread, without being surfeited; they require no preparation; as soon as they are dug up, one may boil and eat them; and it is certain that Europe has more reason to bless the discovery of them, than of all the fruits of both the Indies;* therefore the culture of Potatoes cannot be too much encouraged, nor can I say too much to recommend the use of them; yet there are some observations to be made, by which we shall find,

---

* We owe the discovery of the *Solanum Tuberosum*, which is different from the *Convolvulus caule viride repente*, to Admiral Drake, who discovered them in his first voyage in 1578, in the islands to the westward of the Streights of Magellan, and brought them home with him; but for near a century, they were only cultivated in Ireland, and it is not more than fifty years that they have been cultivated in this country, and but twenty years that they have been common. They did not make that rapid progress in England which might have been expected, although in 1671, it was published in the Philosophical Transactions, No. 90, that they had been of the greatest utility in Ireland, in a dearth which they had suffered the preceding year.

that

that the preference muſt always be given in favour of wheat.

Potatoes are much more bulky than wheat; that is to ſay, there is leſs weight and leſs nouriſhment contained in the ſame ſpace; for this reaſon then, as alſo from their being of a moiſt nature, they do not bear long carriage ſo well, nor are they ſo fit for exportation; and yet they muſt be exported, if there are countries where they are neceſſary, and do not yet grow. It is true, they will thrive in very poor lands, and indeed this might naturally be preſumed from their native ſoil; but they will not grow well in rich land. There is a great difference between thoſe potatoes which grow on our mountains, and thoſe which are produced in the valleys: thoſe in wet lands are bad, and have an acrid diſagreeable taſte, which might render a long uſe of them improper. A ſecond reaſon is, that in dry years their produce is very ſmall. Laſt year, for example, they had not one third of the ordinary crops; and if this were to happen frequently, there would be no reſource; for thoſe of the preceding years cannot be made uſe of, becauſe they will not keep more than a year; except much greater precaution and care be taken of them, than the farmer can attend to, they grow and ſpoil.

The

The independent gentleman, who is fond of them, may preserve some with care till he can dig fresh ones, which he has caused to be planted very early; but I believe the poor cottager must be without them, at least three months, and this is a long time. A third reason is, that they are very soon hurt by the frost; and when they are frozen, they are good for nothing; and the peasant is seldom in a situation to be sure of preserving them in a severe winter; they were almost every where spoiled in those three days of hard frost in January 1776, which, however, did not exceed ten degrees: by this it appears, that potatoes are liable to more inconveniences than wheat; and it must be also remarked, that they require more care.*

It has then been sufficiently demonstrated, that wheat has greatly the advantage even of potatoes: which, however, as Monf. Linguet justly remarks, should be eaten in their natural state, rather than in bread: but I would wish to do all possible justice to M. Parmentier, who has taken great pains to recommend them, and to perfect the art of

---

* If those who have not cultivated them wish to be convinced of this, they should read what is said of them in the *Socrate Rustique*, where their utility is well displayed, and where the culture of them by *Klinck* is so exactly and clearly described, which produced him a very great crop; yet not so large an increase as they are capable of.

baking

baking them in bread: in this respect, we may be more obliged to him than we are at present sensible of, and he merits our grateful acknowledgements; and I think he also deserves much praise for the abilities which he has exerted, and the perseverance he has shewn, in discovering a method of making very fine flour, and very good bread, from potatoes.

I think it would be wrong if he had advised the peasant not to eat them in a natural state, but only in bread; but this, surely, is not his intention; every thing shews, that he only meant to render them more useful; and when an author publishes a work for the public welfare, it would be hard to subject it to severe censure, even if the success of it did not answer his expectation; which, however, is not the case of M. PARMENTIER.

Employed in examining all the farinaceous substances, of which he well knew the qualities and uses, M. PARMENTIER has carried his experiments as far as he could, and has made a discovery, which is not only perfectly safe, but also renders potatoes very useful, since it does not encourage either a monopoly, or laying them up in storehouses, which might occasion a scarcity; but is a mode of making them more generally useful; for if, in

a great

a great plenty of potatoes, the labourer should be satiated with them, or if the servants complain that they have nothing else to eat, in this case, the making them into bread will give that pleasing variety which every one likes; and, as it is very difficult to preserve potatoes from one season to another, the flour prepared in the depth of winter, with one part of good fresh potatoes, would be a very useful resource when they cannot be had in their natural state.

If the idea of making bread from potatoes were as hurtful as Monf. LINGUET affirms, it is not the fault of Monf. PARMENTIER: it has been tried many years. Monf. MUSTEL and Monf. ENGEL, both good citizens, were employed to accomplish the wishes of the peasants in this respect; but their experiments did not completely answer.—Monf. PARMENTIER has discovered that method which was eagerly sought for in vain; and would greatly benefit those countries where they are obliged to use millet, buck-wheat, and maize, if he could also find a method to take off their viscous quality.

To render the common food of any country more salubrious, is adding to the health, strength,

and

and longevity of each individual, and has a greater title to the Civic Crown than saving the life of a single individual; and I heartily congratulate Monſ. FRANCOIS DE NEUCHATEAU, that amiable friend of Monſ. LINGUET, in whom the moſt extenſive knowledge and poetic talents are united with the wiſdom of philoſophy: I heartily congratulate him, I ſay, in having done juſtice to Monſ. PARMENTIER, and in having celebrated, in ſome very fine verſes,* the motive of his works, and the obligations and gratitude due to him. The Authors of the *Journal de Medicine* have alſo been duly ſenſible of the utility of this diſcovery; but they have taken care, at the ſame time, to declare, that however good this bread may be, it is inferior to wheaten bread.

Theſe, Sir, are the principal obſervations which I have made on this Treatiſe of Monſ. LINGUET, and which I thought would be uſeful to thoſe who might, perhaps, be perſuaded by his authority. Men better acquainted with theſe ſubjects than myſelf, might, perhaps, have diſcuſſed them with more preciſion.

I ſhall conclude this letter by remarking, that although the culture of bread-corn may not be

* Le plaiſir de faire le bien
Et le prix de l'homme qui penſe.

detrimental,

detrimental, and although bread is a wholesome food, yet nevertheless, I am perfuaded, as I have already faid, that perhaps (almoft univerfally) too much land is appropriated to the culture of corn; and this error is very general in this country, where, however, it is decreafing.

I do not mean that lefs fhould be reaped: on the contrary, I would wifh to have more plentiful harvefts, becaufe fometimes we have a fcarcity; but I believe, to accomplifh that, (as I have already faid) much lefs fhould be fown; and I am perfuaded, from the beft treatifes on œconomics, from the obfervations communicated to me by perfons fully experienced in this matter, and from the foundeft phyfical principles, that in fowing lefs corn, and putting the remainder of the lands to fome other culture, there would be as much grain reaped, and alfo many other ufeful productions.

It were much to be wifhed, that we could return to the principle of the wife Mr. Kḷiock, and attend clofer to that plain and fimple fyftem of Agriculture, which an experience of thirty years has fully eftablifhed, and which you have defcribed with fo much perfpicuity and elegance.

Your

Your Treatise has been read with the greatest pleasure and eagerness; they have praised, they have admired, they have been enraptured by your wisdom, much more worthy of that name than those sages which were almost adored in Greece. But your work has had the lot of all good books; it has wrought but few changes.—New practices, say they, are plausible, but they are not certain; and therefore they retain the old.

A man, for example, who has thirty arpents of arable land, and a proportionable quantity of meadow, sows, one year with another, twenty arpents, ten to wheat and ten to maize, which is the custom in this country:—For a trial, I should sow but seven; by this method three arpents in each division might be set apart to other uses: Let him sow three arpents of the best land with lucerne, three with sainfoin,* which is perhaps the best of forage, because it will grow in bad land; it will last twelve years, at least as long as lucerne; it affords better nourishment, and for which a little manure, once in three years, is sufficient.†

* Sainfoin is what is called in this country Esparcette [*Onobrychis*.] We call Sainfoin what is elsewhere called Lucerne, [*Medica*.]

† I affirm these facts from my own observations.

Now, it is certain, that these ten arpents would, one year with another, produce sufficient to subsist two horses and four horned cattle; or, which would be much better, seven or eight horned beasts, of which the profit would be more considerable than these six arpents,* if three were sown with wheat and three with maize, indifferently cultivated; and this would be a certain profit, because having his lands better manured, and being enabled to give them one or two more ploughings, his crops will certainly be more plentiful, and subject to less casualties, because the strength of the productions would guard it against accidents, and render it more certain; he would have as much grain; he would be enabled to sell as much or even more, because his cows and the produce of the other three arpents, which I have not yet mentioned, would furnish him with food, which, according to Monsf. Linguet's wish, would serve him instead of bread; he would then be richer and live better, and his substance would be still increased also from the

* Three horses cannot consume annually the first mowing of six arpents sown with lucerne and sainfoin; and as these grasses furnish a second crop more plentiful than the first, lucerne always a third, and sometimes a fourth, and sainfoin often a third; there would remain from these second mowings, and the surplus of the first, sufficient to subsist four horned cattle.

produce

produce of the other three arpents, of which I would wish him to make a kind of kitchen-garden.

This is the method of the wife KLIOCK, the success of which was thought at first to be exaggerated; it has, however, not only borne the test of thirty years experience, but even increased considerably; and what proves it to a demonstration is, that the people of the country followed his example.

Of this garden, a part should be sown with maize, which would enable him to fatten his pigs and poultry; another part with Alsace radishes, either the spring sort, or those which are larger; a good proportion with potatoes; the rest with the yellow carrots, wholesome pulse, light and savory herbs, which are not too tender, but which require as little care as potatoes and cabbages, and which with very little culture will yield more fine plants than will be sufficient for his consumption. The overplus and refuse of his pulse will furnish very excellent food for his sheep and cows.

Carrots are very useful for horses, and the refuse of the cabbages may be used as an exellent manure; and I doubt not but he would live much

better,

better, and be much richer, without more labour; and above all, if to his method he would add some other corrections and amendments to his System of Agriculture, viz.

In the first place, (which has been frequently mentioned already) he should never sow wheat and rye together; for these two grains, tho' of the same kind, do not thrive well in the same soil, nor require the same care in the culture, neither do they ripen at the same time, and never grow so well together as separate; for when the rye grows well, it almost hides the wheat, which, being so much shaded, neither blossoms nor ripens well; therefore, the grain is not so large, nor of so firm a texture, as when the wheat is sown by itself.

The second is, not to let his grass stand too long before it is mown; *first*, because it has been fully proved, that hay is less nourishing to animals, when it comes to feed; and *secondly*, because, as soon as the blossom drops, the plant is nourished wholly from the roots, which impoverishes the land; and *finally*, I wish him to keep a greater number of sheep; but, instead of feeding them in close and hot stables, where they frequently die, he should feed them in inclosures near the house, or in

fields

fields bounded by partitions, six or seven feet high on the north side, and five feet high on the other sides, without any covering, or, if any, only two feet wide on the north side.

The most exact and constant experience, for twenty years, proves, that this is the only method of making them thrive well, and of having fine wool and better-flavoured mutton.

These observations may be depended on as having been made with great precision by Monsf. D'Auberton, one of the most general observers of these days, a gentleman thoroughly acquainted with the nature of those animals, and very exact in his observations, and that in a country farther north, and certainly colder than Lusanne, and which appears to be about the same climate of Zurich and of great part of Switzerland,

This great Physician has proved, that sheep are neither hurt by cold nor by snow, nor rain, but that too great heat hurts them more than any thing else;—an observation which is confirmed by the care they take in Spain to drive them from the plains of Andalusia to the mountains of Old Castile, before the summer heat comes on.

On

On the contrary, in this country, they seem only to fear their being hurt from cold; so they keep them in stables and close places; and by this management, very frequently lose numbers of them.

Thus, sir, have I written a very long letter; and I shall be much flattered, and also be much more certain that I am in the right, if you think as I do.

END OF VOL. I.

www.ingramcontent.com/pod-product-compliance
Lightning Source LLC
Chambersburg PA
CBHW020306240426
43673CB00039B/722